T0214952

Beginning Apache Spark 3

With DataFrame, Spark SQL, Structured Streaming, and Spark Machine Learning Library

Second Edition

Hien Luu

APress®

Beginning Apache Spark 3: With DataFrame, Spark SQL, Structured Streaming, and Spark Machine Learning Library

Hien Luu
SAN JOSE, CA, USA

ISBN-13 (pbk): 978-1-4842-7382-1
https://doi.org/10.1007/978-1-4842-7383-8

ISBN-13 (electronic): 978-1-4842-7383-8

Managing Director, Apress Media LLC: Welmoed Spahr
Acquisitions Editor: Celestin Suresh John
Development Editor: Matthew Moodie
Coordinating Editor: Divya Modi

Cover designed by eStudioCalamar

Cover image designed by Freepik (www.freepik.com)

Distributed to the book trade worldwide by Springer Science+Business Media New York, 1 New York Plaza, New York, NY 10004. Phone 1-800-SPRINGER, fax (201) 348-4505, e-mail orders-ny@springer-sbm.com, or visit www.springeronline.com. Apress Media, LLC is a California LLC and the sole member (owner) is Springer Science + Business Media Finance Inc (SSBM Finance Inc). SSBM Finance Inc is a **Delaware** corporation.

For information on translations, please e-mail booktranslations@springernature.com; for reprint, paperback, or audio rights, please e-mail bookpermissions@springernature.com.

Apress titles may be purchased in bulk for academic, corporate, or promotional use. eBook versions and licenses are also available for most titles. For more information, reference our Print and eBook Bulk Sales web page at http://www.apress.com/bulk-sales.

Any source code or other supplementary material referenced by the author in this book is available to readers on GitHub via the book's product page, located at www.apress.com/9781484273821. For more detailed information, please visit http://www.apress.com/source-code.

Printed on acid-free paper

*I dedicate this book to my wife, Jessica,
and my three boys—Kevin, Steven, and Eric.*

Table of Contents

About the Author

Hien Luu has extensive experience in designing and building big data applications and machine learning infrastructure. He is particularly passionate about the intersection between big data and machine learning. Hien enjoys working with open source software and has contributed to Apache Pig and Azkaban. Teaching is also one of his passions, and he serves as an instructor at the UCSC Silicon Valley Extension school, teaching Apache Spark. He has given presentations at various conferences such as Data+AI Summit, MLOps World, QCon SF, QCon London, Hadoop Summit, and JavaOne.

About the Technical Reviewers

Pramod Singh is a data science manager at Bain & Company. He previously served as a senior machine learning engineer at Walmart Labs and a data science manager at Publicis Sapient in India. He has spent over 11 years in machine learning, deep learning, data engineering, algorithm design, and application development. Pramod has authored four books: *Machine Learning with PySpark* (Apress, 2018), *Learn PySpark* (Apress, 2019), *Learn TensorFlow 2.0* (Apress, 2020), and *Deploy Machine Learning Models to Production* (Apress, 2020). He is also a regular speaker at major conferences such as O'Reilly's Strata Data, GIDS, and other AI conferences. Pramod is an active mentor in the data science community and at various educational institutes. He lives in Gurgaon, India, with his wife and five-year-old son. In his spare time, he enjoys playing guitar, coding, reading, and watching football.

Akshay R. Kulkarni is a renowned AI and machine learning evangelist and thought leader. He has consulted with several Fortune 500 and global enterprises in driving AI and data science–led strategic transformation. Akshay has rich experience in building and scaling AI and machine learning businesses and creating a significant impact. He is currently a manager for Publicis Sapient's core data science and AI team, where he is part of strategy and transformation interventions through AI. He manages high-priority growth initiatives around data science and works on various machine learning and AI engagements by applying state-of-the-art techniques.

Akshay is a Google Developers Expert in machine learning, a published author, and a regular speaker at major AI and data science conferences, including Strata, O'Reilly AI Conf, GIDS. He is also a visiting faculty for some of the top graduate institutes in India.

In 2019, he was featured as one of India's top "40 under 40 Data Scientists".

In his spare time, Akshay enjoys reading, writing, coding, and helping aspiring data scientists. He lives in Bangalore, India, with his family.

Acknowledgments

Writing and completing this book was a team effort that involved many people, and each person played a specific role in helping push this project over the finish line.

First and foremost, I would like to thank my wife for supporting me and giving me space and time to write this book and being OK with skipping some of the house chores.

Second, I would like to thank the technical reviewers, Pramod and Akshay. Their diligence and feedback made this book more useful for readers.

Finally, I would like to thank the ace coordinating editor, Divya Modi, for nudging me and keeping me honest in completing each chapter by the deadline I promised.

Introduction

According to Andrew Ng, AI is the new electricity—powered by big data. It is evident the intersection between big data and AI will grow bigger and stronger as time goes on. Apache Spark was born before the AI revolution. However, it has evolved into an invaluable piece of big data technology to help companies around the world to transform their business with big data and machine learning.

Apache Spark version 3.0 was released in 2020, the same year as Spark's tenth anniversary. Release 3.0 includes many improvements and advancements across the Spark stack. Some of the notable features include 2x performance improvement with adaptive query execution, significant performance improvement and ease of use in panda APIs, and new UI for structured streaming to gain insights into the streaming queries and debug performance-related issues.

There is no better time to learn and gain Apache Spark skills.

CHAPTER 1

Introduction to Apache Spark

There is no better time to learn Apache Spark than now. It has become one of the critical components in the big data stack due to its ease of use, speed, and flexibility. Over the years, it has established itself as the unified engine for multiple workload types, such as big data processing, data analytics, data science, and machine learning. Companies in many industries widely adopt this scalable data processing system, including Facebook, Microsoft, Netflix, and LinkedIn. Moreover, it has steadily improved through each major release.

The more recent version of Apache Spark is 3.0, which was released in June 2020, marking Spark's tenth anniversary as an open source project. This release includes enhancements to many areas of Spark. The notable enhancements are the innovative just-in-time performance optimization techniques to speed up Spark applications and help reduce the time and effort it takes developers to tune their Spark applications.

This chapter provides a high-level overview of Spark, including the core concepts, architecture, and the various components inside the Apache Spark stack.

Overview

Spark is a general distributed data processing engine built for speed, ease of use, and flexibility. The combination of these three properties is what makes Spark so popular and widely adopted in the industry.

The Apache Spark website claims that it can run certain data processing jobs up to 100 times faster than Hadoop MapReduce. In fact, in 2014, Spark won the Daytona GraySort contest, which is an industry benchmark to see how fast a system can sort 100TB of data (1 trillion records). The submission from Databricks claimed Spark could

H. Luu, *Beginning Apache Spark 3*, https://doi.org/10.1007/978-1-4842-7383-8_1

sort 100 TB of data three times faster using ten times fewer resources than the previous world record set by Hadoop MapReduce.

Ease of use has been one of the main focuses of the Spark creators since the inception of the Spark project. It offers over 80 high-level, commonly needed data processing operators to make it easy for developers, data scientists, and analysts to use to build all kinds of interesting data applications. In addition, these operators are available in multiple languages: Scala, Java, Python, and R. Software engineers, data scientists, and data analysts can pick and choose their favorite language to solve large-scale data processing problems with Spark.

In terms of flexibility, Spark offers a single unified data processing stack that can solve multiple types of data processing workloads, including batch applications, interactive queries, machine learning algorithms that require many iterations, and real-time streaming applications to extract actionable insights in near real time. Before the existence of Spark, each of these types of workloads requires a different solution and technology. Now companies can just leverage Spark for all their data processing needs, and it dramatically reduces the operational cost and resources.

The big data ecosystem consists of many pieces of technology, including *Hadoop Distributed File System* (HDFS), a distributed storage engine and cluster management system that efficiently manages a cluster of machines and different file formats to store a large amount of data in binary and columnar formats. Spark integrates well with the big data ecosystem. This is another reason why Spark adoption has been growing at a fast pace.

Another cool thing about Spark is it is open source. Therefore, anyone can download the source code to examine the code, figure out how a certain feature was implemented, and extend its functionalities. In some cases, it can dramatically help reduce the time to debug problems.

History

Spark started as a research project at the University of California, Berkeley, AMPLab in 2009. At that time, the researchers of this project observed the inefficiencies of the Hadoop MapReduce framework in handling interactive and iterative data processing use cases, so they came up with ways to overcome those inefficiencies by introducing ideas like in-memory storage and an efficient way of dealing with fault recovery. Once this research project has proven to be a viable solution that outperforms

MapReduce. It was open sourced in 2010 and became the Apache top-level project in 2013.

Many researchers who worked on this research project founded a company called Databricks, and they raised over $43 million in 2013. Databricks is the primary commercial steward behind Spark. In 2015, IBM announced a major investment in building a Spark technology center to advance Apache Spark by working closely with the open source community and build Spark into the core of the company's analytics and commerce platforms.

Two popular research papers on Spark are "Spark: Cluster Computing with Working Sets" (http://people.csail.mit.edu/matei/papers/2010/hotcloud_spark.pdf) and "Resilient Distributed Datasets: A Fault-Tolerant Abstraction for In-Memory Cluster Computing" (http://people.csail.mit.edu/matei/papers/2012/nsdi_spark.pdf). These papers are well received at academic conferences and provide good foundations for anyone that would like to learn and understand Spark.

Since its inception, the Spark open source project has been a very active project and community. The number of contributors has increased by more than 1000, and there are over 200 thousand Apache Spark meetups. The number of Apache Spark contributors has exceeded the number of contributors of the widely popular Apache Hadoop.

The creators of Spark picked Scala programming language for their project due to the combinations of Scala's conciseness and static typing. Now Spark is considered one of the largest applications written in Scala and its popularity certainly has helped Scala become a mainstream programming language.

Spark Core Concepts and Architecture

Before diving into the details of Spark, it is important to have a high-level understanding of the core concepts and the various core components. This section covers the following.

- Spark clusters

- Resource management system

- Spark applications

- Spark drivers

- Spark executors

Spark Cluster and Resource Management System

Spark is essentially a distributed system designed to process large volumes of data efficiently and quickly. This distributed system is typically deployed onto a collection of machines, known as a Spark cluster. A cluster can be as small as a few machines or as large as thousands of machines. According to the Spark FAQ at `https://spark.apache.org/faq.html`, the world's largest Spark cluster has more than 8000 machines.

Companies rely on a resource management system like Apache YARN or Apache Meso to efficiently and intelligently manage a collection of machines. The two main components in a typical resource management system are cluster manager and worker. The master knows where the slaves are located, how much memory, and the number of CPU cores each one has. One of the main responsibilities of the cluster manager is to orchestrate work by assigning work to workers. Each worker offers resources (memory, CPU, etc.) to the cluster manager and performs the assigned work. An example of this type of work is to launch a particular process and monitor its health. Spark is designed to easily interoperate with these systems. In recent years, most companies adopting big data technologies have a YARN cluster to run MapReduce jobs or other data processing frameworks like Apache Pig or Apache Hive.

Startup companies that fully adopt Spark can just use the out-of-the-box Spark cluster manager to manage a set of machines dedicated to performing data processing using Spark.

Spark Applications

A Spark application consists of two parts. One is the data processing logic expressed using Spark APIs, and the other is the *driver*. Data processing logic can be as simple as a few lines of code to perform a few data processing operations that solve a specific data problem or as complex as training a complicated machine learning model that requires many iterations and runs many hours to complete. A Spark driver is effectively the central coordinator of a Spark application to interact with a cluster manager to figure out which machines to run the data processing logic. For each of those machines, a driver requests a cluster manager to launch a process known as an *executor*.

Another very important job of the Spark driver is managing and distributing Spark tasks onto each executor on behalf of the application. If the data processing logic requires the Spark driver to collect the computed results to present to a user, it coordinates with each Spark executor to collect the computed result and merge them together before presenting them to the user. A Spark driver performs tasks through a component called `SparkSession`.

Spark Drivers and Executors

Each Spark executor is a JVM process and is dedicated to a specific Spark application. The life span of a Spark executor is the duration of a Spark application, which could be minutes or days. There was a conscious design decision not to share a Spark executor between different multiple Spark applications. This has the benefit of isolating each application from each other. Still, it is not easy to share data between different applications without writing that data to an external storage system like HDFS.

In short, Spark employs a master/slave architecture, where the driver is the master, and the executor is the slave. Each of these components runs as an independent process on a Spark cluster. A Spark application consists of one driver and one or more executors. Playing the slave role, a Spark executor does what is being told, which is to execute the data processing logic in the form of tasks. Each task is executed on a separate CPU core. This is how Spark parallelly processes data to speed things up. In addition, each Spark executor is responsible for caching a portion of the data in memory and/or on disk when it is told to do so by the application logic.

When launching a Spark application, you can specify the number of executors the application needs, and the amount of memory and the number of CPU cores each executor should have.

Figure 1-1 shows interactions between a Spark application and cluster manager.

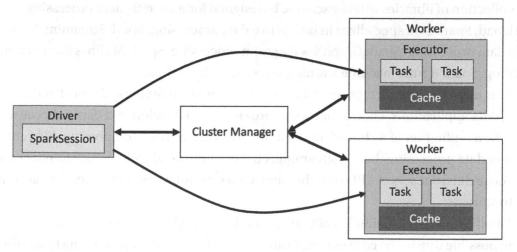

Figure 1-1. *Interactions between a Spark application and the cluster manager*

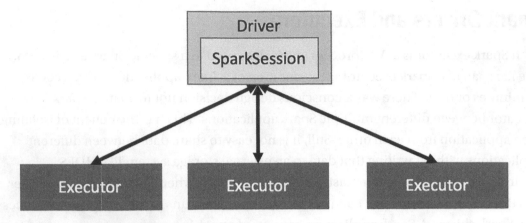

Figure 1-2. *A Spark cluster that consists of one driver and three executors*

Spark Unified Stack

Unlike its predecessors, Spark provides a unified data processing engine known as the
Spark stack. Like other well-designed systems, this stack is built on a strong foundation
called Spark Core, which provides all the necessary functionality to manage and run
distributed applications like scheduling, coordination, and handling fault tolerance.
In addition, it provides a powerful and generic programming abstraction for data
processing called *resilient distributed datasets* (RDDs). On top of this strong foundation
is a collection of libraries where each one is designed for a specific data processing
workload. Spark SQL specializes in interactive data processing. Spark Streaming is real-
time data processing. Spark GraphX is for graph processing. Spark MLlib is for machine
learning. Spark R runs machine learning tasks using the R shell.

This unified engine brings several important benefits to building the next generation
of big data applications. First, applications are simpler to develop and deploy because
they use a unified set of APIs and run on a single engine. Second, combining different
types of data processing (batch, streaming, etc.) is far more efficient because Spark can
run those different sets of APIs over the same data without writing the intermediate data
out to storage.

Finally, the most exciting benefit is that Spark enables brand-new applications
made possible due to the ease of composing different sets of data processing types; for
example, running interactive queries on the results of machine learning predictions
of real-time data streams. An analogy that everyone can relate to is a smartphone,

consisting of a powerful camera, cellphone, and GPS device. By combining the functions of these components, smartphones enable innovative applications like Waze, a traffic and navigation application.

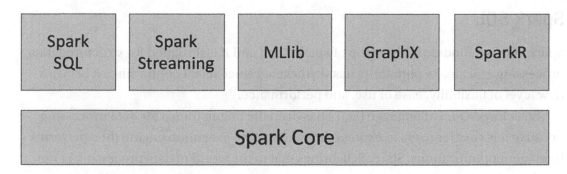

Figure 1-3. *Spark unified stack*

Spark Core

Spark Core is the bedrock of the Spark distributed data processing engine. It consists of an RDD, a distributed computing infrastructure and programming abstraction.

The distributed computing infrastructure is responsible for distributing, coordinating, and scheduling computing tasks across many machines in the cluster. This enables the ability to perform parallel data processing of large volumes of data efficiently and quickly on a large cluster of machines. Two other important responsibilities of the distributed computing infrastructure are handling computing task failures and the efficient way of moving data across machines, known as data shuffling. Advanced Spark users should have intimate knowledge of Spark distributed computing infrastructure to effectively design high-performance Spark applications.

The RDD key programming abstraction is something that every Spark user should learn and effectively use the various provided APIs. An RDD is a fault-tolerant collection of objects partitioned across a cluster that can be manipulated in parallel. Essentially it provides a set of APIs for Spark application developers to easily and efficiently perform large-scale data processing without worrying where data resides on the cluster and machine failures. The RDD APIs are exposed to multiple programming languages, including Scala, Java, and Python. They allow users to pass local functions to run on the cluster, which is very powerful and unique. RDDs are covered in detail in a later chapter.

The rest of the components in the Spark stack are designed to run on top of Spark Core. Therefore, any improvement or optimization done in the Spark Core between versions of Spark is automatically available to the other components.

Spark SQL

Spark SQL is a module built on top of Spark Core, and it is designed for structured data processing at scale. Its popularity has skyrocketed since its inception since it brings a new level of flexibility, ease of use, and performance.

Structured Query Language (SQL) has been the lingua franca for data processing because it is easy for users to express their intent. The execution engine then performs intelligent optimizations. Spark SQL brings that to the world of data processing at the petabytes level. Spark users now can issue SQL queries to perform data processing or use the high-level abstraction exposed through the DataFrame API. A DataFrame is effectively a distributed collection of data organized into named columns. This is not a new idea. It is inspired by data frames in R and Python. An easier way to think about a DataFrame is that it is conceptually equivalent to a table in a relational database.

Behind the scenes, the Spark SQL Catalyst optimizer performs optimizations commonly done in many analytical database engines.

Another Spark SQL feature that elevates Spark's flexibility is the ability to read and write data to and from various structured formats and storage systems, such as JavaScript Object Notation (JSON), comma-separated values (CSV), Parquet or ORC files, relational databases, Hive, and others.

According to the 2021 Spark survey, Spark SQL was the fastest-growing component. This makes sense because Spark SQL enables a wider audience beyond "big data" engineers to leverage the power of distributed data processing—that is, data analysts or anyone familiar with SQL.

The motto for Spark SQL is to write less code, read less data, and the optimizer does the hard work.

Spark Structured Streaming

It has been said that "data in motion has equal or greater value than historical data." The ability to process data as they arrive has become a competitive advantage for many companies in highly competitive industries. The Spark Structured Streaming module enables the ability to process real-time streaming data from various data sources in a high-throughput and fault-tolerant manner. Data can be ingested from sources like Kafka, Flume, Kinesis, Twitter, HDFS, or TCP socket.

Spark's main abstraction for processing streaming data is a discretized stream (DStream), which implements an incremental stream processing model by splitting the input data into small batches (based on a time interval) that can regularly combine the current processing state to produce new results.

Stream processing sometimes involves joining with data at rest, and Spark makes it very easy. In other words, combining batch and interactive queries with stream processing can be easily done in Spark due to the unified Spark stack.

A new scalable and fault-tolerant stream processing engine called Structured Streaming was introduced in Spark version 2.1. This engine further simplifies stream processing app developers' lives by treating streaming computation the same way as you express a batch computation on static data. This new engine automatically executes the stream processing logic incrementally and continuously and produces the result as new streaming data arrives. Another unique feature in the Structured Streaming engine is the guarantee of end-to-end exactly-once support, which makes "big data" engineer's life much easier than before in terms of saving data to a storage system like a relational database or a NoSQL database.

As this new engine matures, it enables a new class of stream processing applications that are easy to develop and maintain.

According to Reynold Xin, Databricks' chief architect, the simplest way to perform streaming analytics is not having to reason about streaming.

Spark MLlib

MLlib is Spark's machine learning library. It provides more than 50 common machine learning algorithms and abstractions for managing and simplifying model-building tasks, such as featurization, a pipeline for constructing, an evaluating and tuning model, and the persistence of models to help move models from development to production.

Starting with Spark 2.0 version, the MLlib APIs are based on DataFrames to take advantage of the user-friendliness and many optimizations provided by the Catalyst and Tungsten components in the Spark SQL engine.

Machine learning algorithms are iterative, meaning they run through many iterations until the desired objective is achieved. Spark makes it extremely easy to implement those algorithms and run them in a scalable manner through a cluster of machines. Commonly used machine learning algorithms such as classification, regression, clustering, and collaborative filtering are available out of the box for data scientists and engineers to use.

Spark GraphX

Graph processing operates on a data structure consisting of vertices and edges connecting them. A graph data structure is often used to represent real-life networks of interconnected entities, including professional social networks on LinkedIn, a network of connected web pages on the Internet, and so on. Spark GraphX is a library that enables graph-parallel computations by providing an abstraction of a directed multigraph with properties attached to each vertex and edge. GraphX includes a collection of common graph processing algorithms, including page ranks, connected components, shortest paths, and others.

SparkR

SparkR is an R package that provides a lightweight frontend to use Apache Spark. R is a popular statistical programming language that supports data processing and machine learning tasks. However, R was not designed to handle large datasets that cannot fit on a single machine. SparkR leverages Spark's distributed computing engine to enable large-scale data analysis using familiar R shell and popular APIs that many data scientists love.

Apache Spark 3.0

The 3.0 release has new features and enhancements to most of the components in the Spark stack. However, about 60% of the enhancements went into Spark SQL and Spark Core components. Query performance optimization was one of the major themes in Spark 3.0, so the bulk of the focus and development was in the Spark SQL component. Based on the TPC-DS 30 TB benchmark done by Databricks, Spark 3.0 is roughly two times faster than Spark 2.4. This section highlights a few notable features that are related to performance optimization.

Adaptive Query Execution Framework

As the name suggests, the query execution framework adapts the execution plan at runtime based on the most recent statistics about data size, the number of partitions, and so forth. As a result, Spark can dynamically switch join strategies, automatically optimize skew joins, and adjust the number of partitions. All these intelligent optimizations lead to improving the query performance of Spark applications.

Dynamic Partition Pruning (DPP)

The primary idea behind DPP is simple, which is to avoid reading unnecessary data. It is designed specifically for use cases when querying data using joins against fact tables and dimension tables in a star schema scheme. It can dramatically improve the join performance by reducing the number of rows in the fact table that need to join with the dimension tables based on the given filtering conditions. Based on a TPC-DS benchmark, this optimization technique can speed up the performance of 60% of the queries in the range of 2x to 18x.

Accelerator-aware Scheduler

More and more Spark users are leveraging Spark for both big data processing and machine learning workload. The latter type of workload often needs GPU to speed up the machine learning model training process. This enhancement enables Spark users to describe and request GPU resources for their complex workloads that involve machine learning.

Apache Spark Applications

Spark is a versatile, fast, and scalable data processing engine. It was designed to be a general engine since the beginning days and has proven that it can be used to solve many use cases. As a result, many companies in various industries are using Spark to solve many real-life use cases. The following is a small list of applications that were developed using Spark.

- Customer intelligence application

- Data warehouse solutions

- Real-time streaming solutions

- Recommendation engines
- Log processing
- User-facing services
- Fraud detection

Spark Example Applications

In the world of big data processing, the canonical example application is the word count application. This tradition started with the introduction of the MapReduce framework. Since then, every big data processing technology-related book must follow this unwritten tradition by including this canonical example. The problem space in the word count example application is easy for everyone to understand since all it does is count how many times a particular word appears in each set of documents, whether that is a chapter of a book or hundreds of terabytes of web pages from the Internet.

Listing 1-1 is a word count example application in Spark in the Scala language.

Listing 1-1. The Word Count Spark Example Application Written in Scala Language

```
val textFiles = sc.textFile("hdfs://<folder>")
val words = textFiles.flatMap(line => line.split(" "))
val wordTuples = words.map(word => (word, 1))
val wordCounts = wordTuples.reduceByKey(_ + _)
wordCounts.saveAsTextFile("hdfs://<outoupt folder>")
```

A lot is going on behind these five lines of code. The first line is responsible for reading the text files under the specified folder. The second line iterates through each line in each of the files, then each line is tokenized into an array of words and finally flattens each array into one word per line. The third line attaches a count of 1 to each word to count the number of words across all documents. The fourth line performs the summation of the count of each word. Finally, the last line saves the result in the specified folder. Hopefully, this gives you a general sense of the ease of use of Spark to perform data processing. Future chapters go into more detail about what each of those lines of code does.

Apache Spark Ecosystem

In the realm of big data, innovation doesn't stand still. As time goes on, the best practices and architectures emerge. The Spark ecosystem is expanding and evolving to address some of the emerging needs in data lakes, helping data scientists be more productive at interacting with the vast amount of data and speeding up the machine learning development life cycle. This section highlights a few of the exciting and recent innovations in the Spark ecosystem.

Delta Lake

At this point, most companies recognize the value of data and have some form of strategy to ingest, store, process, and extract insights from their data. The idea behind Delta Lake is to leverage a distributed storage solution to store both structured and unstructured data for various data consumers such as data scientists, data engineers, and business analysts. To ensure the data in Delta Lake is usable, there must be oversights in the data catalog, data discovery, data quality, access control, and data consistency semantics. Data consistency semantics presents many challenges, and companies have invented tricks or "Band-Aid" solutions.

Delta Lake is an open source solution for data consistency semantics that provides an open data storage format with transactional guarantees and schema enforcement and evolution support. Delta Lake is further discussed later.

Koalas

For years, data scientists have been using the Python pandas library to perform data manipulation in their machine learning–related tasks. The pandas library (`https://pandas.pydata.org`) is a "fast, powerful, flexible and easy to use open source data analysis and manipulation tool built on top of Python programming language." pandas is widely popular and has become the de facto library due to its powerful and flexible abstraction called a DataFrame for data manipulation. However, pandas is designed to run on a single machine only. To perform parallel computing in Python, you can explore an open source project called Dask (`https://docs.dask.org`).

Koalas marries the best of both worlds, the powerful and flexible DataFrame abstraction and Spark's distributed data processing engine by implementing the pandas DataFrame API on top of Apache Spark.

This innovation enables data scientists to leverage their pandas knowledge to interact with much bigger datasets than in the past.

Koalas version 1.0 was released in June 2020 with 80% coverage of the pandas APIs. Koalas aims to enable data science projects to leverage large datasets instead of being blocked by them.

MLflow

The field of machine learning has been around a long time. Recently, it has become more approachable due to advancements in algorithms, ease of access to a large collection of useful datasets such as images and a large corpus of text, and the availability of educational resources. However, applying machine learning to business problems has proven to be a challenge because it is more of a software engineering problem to manage the machine learning life cycle.

MLflow is an open source project. It was conceived in 2018 to provide a platform to help with managing the machine learning life cycle. It consists of the following components to address the various needs in each step of the life cycle.

- **Tracking** records and compares machine learning experiments.

- **Projects** provides a consistent format of organizing machine learning projects to share and reproduce machine learning models easily.

- **Models** provides a standardized format to package machine learning models, a consistent API for working with machine learning models, such as loading and deploying them.

- **Registry** is a model store that hosts machine learning models and tracks their lineage, version, and deployment state transitions.

Summary

- Apache Spark has certainly produced many sparks since its inception. It has created much excitement and opportunities in the world of big data. And more importantly, it allows you to create many new and innovative big data applications to solve a diverse set of data processing problems of data applications.

- The three important properties of Spark to note are ease of use, speed, and flexibility.

- The Spark distributed computing infrastructure employs a master and slave architecture. Each Spark application consists of a driver and one or more executors to process the data in parallel. Parallelism is the key enabler to process massive amounts of data in a short amount of time.

- Spark provides a unified scalable and distributed data processing engine that can be used for batch processing, interactive and exploratory data processing, real-time stream processing, building machine learning models and predictions, and graph processing.

- Spark applications can be written in multiple programming languages, including Scala, Java, Python, or R.

CHAPTER 2

Working with Apache Spark

When it comes to working with Spark or building Spark applications, there are many options. This chapter describes the three common options, including using Spark shell, submitting a Spark application from the command line, and using a hosted cloud platform called Databricks. The last part of this chapter is geared toward software engineers who want to set up Apache Spark source code on a local machine to study Spark source code and learn how certain features were implemented.

Downloading and Installation

To learn or experiment with Spark, it is convenient to have it installed locally on your computer. This way, you can easily try out certain features or test your data processing logic with small datasets. Having Spark locally installed on your laptop lets you learn it from anywhere, including your comfortable living room, the beach, or at a bar in Mexico.

Spark is written in Scala. It is packaged so that it can run on both Windows and UNIX-like systems (e.g., Linux, macOS). To run Spark locally, all that is needed is Java installed on your computer.

To set up a multitenant Spark production cluster requires a lot more information and resources, which are beyond the scope of this book.

Downloading Spark

The Download section of the Apache Spark website (http://spark.apache.org/downloads.html) has detailed instructions for downloading the pre-packaged Spark binary file. At the time of writing this book, the latest version is 3.1.1. In terms of package type, choose the one with the latest version of Hadoop. Figure 2-1 shows the various

© Hien Luu 2021
H. Luu, *Beginning Apache Spark 3*, https://doi.org/10.1007/978-1-4842-7383-8_2

options for downloading Spark. The easiest way is to download the pre-packaged binary file because it contains the necessary JAR files to run Spark on your computer. Clicking the link on line item 3 triggers the binary file download. There is a way to manually build the Spark binary from source code. The instructions on how to do that are covered later in the chapter.

Download Apache Spark™

1. Choose a Spark release: 3.1.1 (Mar 02 2021) ⬦

2. Choose a package type: Pre-built for Apache Hadoop 2.7   ⬦

3. Download Spark: spark-3.1.1-bin-hadoop2.7.tgz

4. Verify this release using the 3.1.1 signatures, checksums and project release KEYS.

Note that, Spark 2.x is pre-built with Scala 2.11 except version 2.4.2, which is pre-built with Scala 2.12. Spark 3.0+ is pre-built with Scala 2.12.

Figure 2-1. *Apache Spark download options*

Installing Spark

Once the file is successfully downloaded onto your computer, the next step is to uncompress it. The `spark-3.1.1-bin-hadoop2.7.tgz` file is in a GZIP compressed tar archive file, so you need to use the right tool to uncompress it.

For Linux or macOs computers, the `tar` command should already exist. So run the following command to uncompress the downloaded file.

```
tar xvf spark-3.1.1-bin-hadoop2.7.tgz
```

For Windows computers, you can use either the WinZip or 7-zip tool to unzip the downloaded file.

Once the uncompression is successfully finished, there should be a directory called spark-3.1.1-bin-hadoop2.7. From here on, this directory is referred to as the Spark directory.

Note   If a different version of Spark is downloaded, the directory name is slightly different.

There are about a dozen directories under the `spark-3.1.1-bin-hadoop2.7` directory. Table 2-1 describes the ones that are good to know.

Table 2-1. *The Subdirectories in spark-3.1.1-bin-hadoop2.7*

Name	Description
bin	Contains the various executable files to bring up Spark shell in Scala or Python, submit Spark applications, run Spark examples
data	Contains small sample data files for various Spark examples
examples	Contains both the source code and binary file for all Spark examples
jars	Contains the necessary binaries that are needed to run Spark
sbin	Contains the executable files to manage Spark cluster

The next step is to test out the installation by bringing up the Spark shell.

Spark shell is like a Unix shell. It provides an interactive environment to easily learn Spark and analyze data. Most Spark applications are developed using either Python or Scala programming language. Spark shell is available for both of those languages. If you are a data scientist and Python is your cup of tea, you will not feel left out. The following section shows how to bring up Spark Scala and Spark Python shell.

Note   Scala is a Java JVM-based language, and thus, it is easy to leverage existing Java libraries in Scala applications.

Spark Scala Shell

To start up the Spark Scala shell, enter the **./bin/spark-shell** command in the Spark directory. After a few seconds, you should see something similar to Figure 2-2.

```
Setting default log level to "WARN".
To adjust logging level use sc.setLogLevel(newLevel). For SparkR, use setLogLevel(newLevel).
Spark context Web UI available at http://192.168.0.22:4040
Spark context available as 'sc' (master = local[*], app id = local-1615662492205).
Spark session available as 'spark'.
Welcome to
      ____              __
     / __/__  ___ _____/ /__
    _\ \/ _ \/ _ `/ __/  '_/
   /___/ .__/\_,_/_/ /_/\_\   version 3.1.1
      /_/

Using Scala version 2.12.10 (OpenJDK 64-Bit Server VM, Java 11.0.6)
Type in expressions to have them evaluated.
Type :help for more information.

scala> ▮
```

Figure 2-2. *Scala Spark shell output*

To exit the Scala Spark shell, type **:quit** or **:q**.

Note   Java version 11 or higher is preferred to run the Spark Scala shell.

Spark Python Shell

To start up the Spark Python shell, enter the **./bin/pyspark** command in the Spark directory. After a few seconds, you should see something similar to Figure 2-3.

```
Using Spark's default log4j profile: org/apache/spark/log4j-defaults.properties
Setting default log level to "WARN".
To adjust logging level use sc.setLogLevel(newLevel). For SparkR, use setLogLevel(newLevel).
Welcome to
      ____              __
     / __/__  ___ _____/ /__
    _\ \/ _ \/ _ `/ __/  '_/
   /__ / .__/\_,_/_/ /_/\_\   version 3.1.1
      /_/

Using Python version 3.7.6 (default, Jan  8 2020 13:42:34)
Spark context Web UI available at http://192.168.0.22:4040
Spark context available as 'sc' (master = local[*], app id = local-1615663017956).
SparkSession available as 'spark'.
>>> ▮
```

Figure 2-3. *Output of Python Spark shell*

To exit the Python Spark shell, enter **ctrl-d**.

Note   Spark Python shell requires Python 3.7.x or higher.

The Spark Scala shell and the Spark Python shell are extensions of Scala REPL and Python REPL, respectively. REPL is an acronym for read-eval-print loop. It is an interactive computer programming environment that takes user inputs, evaluates them, and returns the result to the user. Once a line of code is entered, the REPL immediately provides feedback on whether there is a syntactic error. If there aren't any syntax errors, it evaluates them. If there is any output, it is displayed in the shell. The interactive and immediate feedback environment allows developers to be very productive by bypassing the code compilation step in the normal software development process.

To learn Spark, Spark shell is a very convenient tool to use on your local computer anytime and anywhere. It doesn't have any external dependencies other than the data files you process need to reside on your computer. However, if you have an Internet connection, it is possible to access those remote data files, but it will be slow.

The remaining chapters of this book use the Spark Scala shell.

Having Fun with the Spark Scala Shell

This section provides information about Scala Spark shell and a set of useful commands to know to be effective and productive at using it for exploratory data analysis or building Spark applications interactively.

The `./bin/spark-shell` command effectively starts a Spark application and provides an environment where you can interactively call Spark Scala APIs to easily perform exploratory data processing. Since Spark Scala shell is an extension of Scala REPL, it is a great way to use it to learn Scala and Spark at the same time.

Useful Spark Scala Shell Command and Tips

Once a Spark Scala shell is started, it puts you in an interactive environment to enter shell commands and Scala code. This section covers various useful commands and a few tips on working in the shell.

Once inside the Spark Shell, type the following to bring a complete list of available commands.

```
scala>  :help
```

The output of this command is shown in Figure 2-4.

```
scala> :help
All commands can be abbreviated, e.g., :he instead of :help.
:completions <string>     output completions for the given string
:edit <id>|<line>         edit history
:help [command]           print this summary or command-specific help
:history [num]            show the history (optional num is commands to show)
:h? <string>              search the history
:imports [name name ...]  show import history, identifying sources of names
:implicits [-v]           show the implicits in scope
:javap <path|class>       disassemble a file or class name
:line <id>|<line>         place line(s) at the end of history
:load <path>              interpret lines in a file
:paste [-raw] [path]      enter paste mode or paste a file
:power                    enable power user mode
:quit                     exit the interpreter
:replay [options]         reset the repl and replay all previous commands
:require <path>           add a jar to the classpath
:reset [options]          reset the repl to its initial state, forgetting all session entries
:save <path>              save replayable session to a file
:sh <command line>        run a shell command (result is implicitly => List[String])
:settings <options>       update compiler options, if possible; see reset
:silent                   disable/enable automatic printing of results
:type [-v] <expr>         display the type of an expression without evaluating it
:kind [-v] <type>         display the kind of a type. see also :help kind
:warnings                 show the suppressed warnings from the most recent line which had any
```

Figure 2-4. *List of available shell commands*

Some commands are used more often than others because of their usefulness. Table 2-2 describes the commonly used commands.

Table 2-2. *Useful Spark Shell Commands*

Name	Description
:history	This command displays what was entered during the previous Spark shell session as well as the current session. It is useful for copying purposes.
:load	Load and execute the code in the provided file. This is particularly useful when the data processing logic is long. It is a bit easier to keep track of the logic in a file.
:reset	After experimenting with the various Scala or Spark APIs for a while, you may lose track of the value of various variables. This command resets the shell to a clean state to make it easy to reason.
:silent	This is for an advanced user who is a bit tired at looking at the output of each Scala or Spark APIs that were entered in the shell. To re-enable the output, simply type :silent again.
:quit	This is a self-explanatory command but useful to know. Often, people try to quit the shell by entering :exit, which doesn't work.
:type	Display the type of a variable. :type <variable name>

In addition to these commands, a helpful feature for improving developer productivity is the code completion feature. Like popular integrated development environments (IDEs) like Eclipse or IntelliJ, the code completion feature helps developers explore the possible options and reduce typing errors.

Inside the shell, type **spa** and then hit the Tab key. The environment adds characters to transform "spa" to "spark". In addition, it shows possible matches for Spark (see Figure 2-5).

```
scala> spa <tab>
```

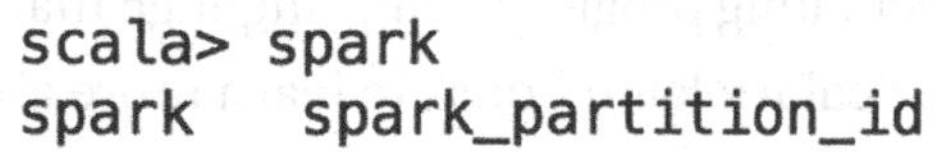

```
scala> spark
spark    spark_partition_id
```

Figure 2-5. *Tab completion output of spa*

In addition to completing the name of a partially entered word, the tab completion can show an object's available member variables and functions.

In the shell, type **spark**, and then hit the Tab key. This displays a list of available member variables and functions of the Scala object represented by the spark variable (see Figure 2-6).

```
scala> spark.
baseRelationToDataFrame   createDataFrame   executeCommand    newSession    sessionState   sqlContext   time
catalog                   createDataset     experimental      range         sharedState    stop         udf
close                     emptyDataFrame    implicits         read          sparkContext   streams      version
conf                      emptyDataset      listenerManager   readStream    sql            table
```

Figure 2-6. *List of available member variables and functions of object called "spark"*

The :history command displays the previously entered commands or lines of code. This suggests that the Spark shell maintains a record of what was entered. One way to quickly display or recall what was entered recently is by pressing the up arrow key. Once you scroll up to the line you want to execute, simply hit Enter to execute it.

Basic Interactions with Scala and Spark

The preceding section covered the basics of navigating the Spark shell; this section introduces a few fundamental ways of working with Scala and Spark in Spark shell. This fundamental knowledge will be really helpful in future chapters as you dive much deeper into topics like Spark DataFrame and Spark SQL.

Basic Interactions with Scala

Let's start with Scala in the Spark Scala shell, which provides a full-blown environment for learning Scala. Think of Spark Scala shell as a Scala application with an empty body, and this is where you come in. You fill this empty body with Scala functions and logic for your application. This section intends to demonstrate a few simple Scala examples in Spark shell. Scala is a fascinating programming language that is powerful, concise, and elegant. Please refer to Scala-related books to learn more about this programming language.

The canonical example for learning any programming language is the "Hello World" example, which entails printing out a message. So let's do that. Enter the following line in the Spark Scala shell; the output should look like Figure 2-7.

```scala
scala> println("Hello from Spark Scala shell")
```

```
scala> println("Hello from Spark Scala shell")
Hello from Spark Scala shell

scala> █
```

Figure 2-7. *Output of the Hello World example command*

The next example defines an array of ages and prints those element values out in the Spark shell. In addition, this example illustrates the code completion feature mentioned in the previous section.

To define an array of ages and assign it to an immutable variable, enter the following into the Spark shell. Figure 2-8 shows the evaluation output.

```
scala> val ages = Array(20, 50, 35, 41)
```

```
scala> val ages = Array(20, 50, 35, 41)
ages: Array[Int] = Array(20, 50, 35, 41)
```

Figure 2-8. *Output of defining an array of ages*

Now you can refer to the ages variable in the following line of code. Let's pretend that you can't exactly remember a function name in the Array class to iterate through the elements in the array, but you know it starts with "fo". You can enter the following and hit the tab to see how Spark shell can help.

```
scala> ages.fo
```

After you press the Tab key, Spark shell displays what's shown in Figure 2-9.

```
scala> ages.fo
fold     foldLeft    foldRight    forall    foreach    formatted
```

Figure 2-9. *Output of code completion*

Aha! You need the foreach function to iterate through the elements in the array. Let's use it to print the ages.

```
scala> ages.foreach(println)
```

Figure 2-10 shows the expected output.

```
[scala> ages.foreach(println)
20
50
35
41
```

Figure 2-10. *Output from printing out the ages*

The previous code statement may look a bit cryptic for those new to Scala; however, you can intuitively guess what it does. As the foreach function iterates through each element in the "ages" array, it passes that element to the println function to print the value out to the console. This style is used quite a bit in the coming chapters.

The last example in this section defines a Scala function to determine whether the age is an odd number or even number; it is then used to find the odd number ages in the array.

```
scala> def isOddAge(age:Int) : Boolean = {
  (age % 2) == 1
}
```

If you come from a Java programming background, this function signature may look strange, but it is not too difficult to decipher what it does. Notice the function doesn't use the return keyword to return the value of the expression in its body. In Scala, it is not necessary to add the return keyword. The output of the last statement in a function body is returned to the caller (if that function was defined to return a value). Figure 2-11 shows the output from the Spark shell.

```
[scala> def isOddAge(age:Int) : Boolean = {
[     |    (age % 2) == 1
[     | }
 isOddAge: (age: Int)Boolean
```

Figure 2-11. *If there are syntax errors, Spark shell returns the function signature*

To figure out the odd number ages in the ages array, let's leverage the filter function in the Array class.

```
scala> ages.filter(age => isOddAge(age)).foreach(println)
```

This line of code does the filtering and then iterates through the result to print out the odd ages. It is a common practice in Scala to use function chaining to make the code concise. Figure 2-12 shows the output from Spark shell.

```
[scala> ages.filter(age => isOddAge(age)).foreach(println)
35
41
```

Figure 2-12. *The output from filtering and printing out only ages that are odd numbers*

Now let's try out the :type shell command on a Scala variable and function defined earlier. This command comes in handy once you have used Spark shell for a while and lost track of the data type of a certain variable or the return type of a function. Figure 2-13 shows examples of the :type command.

```
[scala> :type ages
Array[Int]

[scala> :type isOddAge(100)
Boolean
```

Figure 2-13. *Output of :type command*

To learn Spark, it is not necessary to master the Scala programming language. However, one must be comfortable with knowing and working with the basics of Scala. A good resource for learning just enough Scala to learn Spark is at https://github.com/deanwampler/JustEnoughScalaForSpark. This resource was presented at various Spark-related conferences.

Spark UI and Basic Interactions with Spark

In the previous section, I mentioned Spark shell is a Scala application. That is only partially true. The Spark shell is a Spark application written in Scala. When the Spark shell is started, a few things are initialized and set up for you to use, including Spark UI and a few important variables.

Spark UI

If you go back and carefully examine the Spark shell output in either Figure 2-2 or Figure 2-3, you see a line that looks something like the following. (The URL may be a bit different for your Spark shell.)

The SparkContext Web UI is available at http://<ip>:4040.

If you point your browser to that URL in your Spark shell, it displays something like what's shown in Figure 2-14.

Figure 2-14. *Spark UI*

The Spark UI is a web application designed to help with monitoring and debugging Spark applications. It contains detailed runtime information and various resource consumptions of a Spark application. The runtime includes various metrics that are tremendously helpful in diagnosing performance issues in your Spark applications. One thing to note is that the Spark UI is only available while a Spark application is running.

The navigation bar at the top of the Spark UI contains links to the various tabs, including Jobs, Stages, Storage, Environment, Executors, and SQL. I briefly cover the Environment and Executors tabs and describe the remaining tabs in later chapters.

The Environment tab contains static information about the environment that a Spark application is running in. This includes runtime information, spark properties, system properties, and classpath entries. Table 2-3 describes each of those areas.

Table 2-3. *Sections in the Environment Tab*

Name	Description
Runtime Information	Contains the locations and versions of the various components that Spark depends on, including Java and Scala.
Spark Properties	This area contains the basic and advanced properties that are configured in a Spark application. The basic properties include the basic information about an application like application id, name, and so on. The advanced properties are meant to turn on or off certain Spark features or tweak them in certain ways that are best for a particular application. See the resource at `https://spark.apache.org/docs/latest/configuration.html` for a comprehensive list of configurable properties.
Resource Profiles	Information about the number of CPUs and the amount of memory in the Spark cluster.
Hadoop Properties	The various Hadoop and Hadoop File System properties.
System Properties	These properties are mainly at the OS and Java level, not Spark specific.
Classpath Entries	Contains a list of classpaths and jar files that are used in a Spark application.

The Executors tab contains the summary and breakdown information for each of the executors supporting a Spark application. This information includes the capacity of certain resources and how much is being used in each executor. The resources include memory, disk, and CPU. The Summary section provides a bird's-eye view of the resource consumption across all the executors in a Spark application. Figure 2-15 shows more of the details.

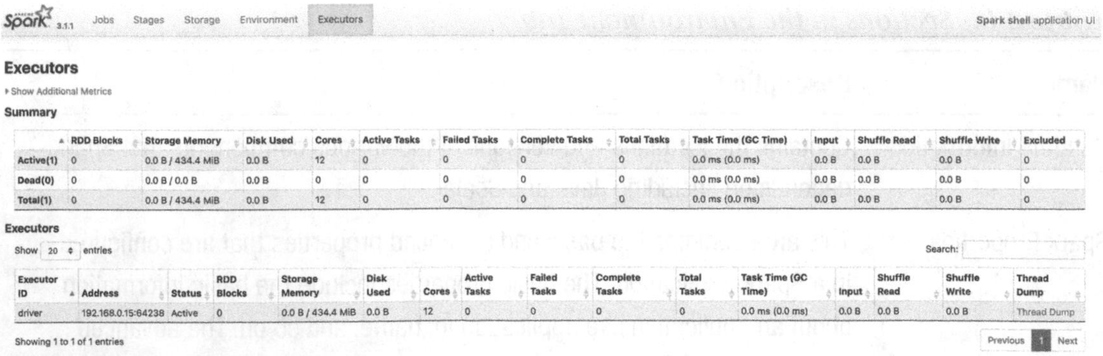

Figure 2-15. *Executor tab of a Spark application that uses only a single executor*

You revisit Spark UI in a later chapter.

Basic Interactions with Spark

Once a Spark shell is successfully started, an important variable called spark is initialized and ready to be used. The spark variable represents an instance of a SparkSession class. Let's use the :type command to verify this.

```
scala>:type spark
```

And the Spark shell displays its type in Figure 2-16.

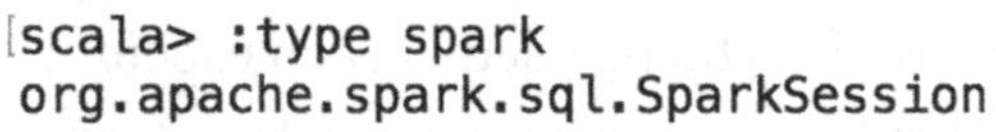

```
[scala> :type spark
org.apache.spark.sql.SparkSession
```

Figure 2-16. *Showing the type of "spark" variable*

The SparkSession class was introduced in Spark 2.0 to provide a single entry point to interact with underlying Spark functionalities. This class has APIs for reading unstructured and structured data in text and binary formats, such as JSON, CSV, Parquet, ORC, and so on. In addition, the SparkSession component provides a facility to retrieve and set Spark configurations.

Let's start interacting with the spark variable in Spark shell to print out a few useful pieces of information, such as the version and existing configurations. From the Spark shell, type the following code to print the Spark version. Figure 2-17 shows the output.

```scala
scala> spark.version
```

```
[scala> spark.version
res0: String = 3.1.1
```

Figure 2-17. *Spark version output*

To be a little more formal, you can use the println function covered in the previous section to print out the Spark version and output shown in Figure 2-18.

```scala
scala> println("Spark version: " + spark.version)
```

```
[scala> println("Spark vesion:" + spark.version)
Spark vesion:3.1.1
```

Figure 2-18. *Display Spark version using println function*

To see the default configurations in the Spark shell, you access the conf variable of spark. Here is the code to display the default configurations, and the output is shown in Figure 2-19.

```scala
scala> spark.conf.getAll.foreach(println)
```

```
[scala> spark.conf.getAll.foreach(println)
(spark.driver.host,192.168.1.73)
(spark.driver.port,54812)
(hive.metastore.warehouse.dir,file:/Users/spark-user/spark-2.1.1-bin-hadoop2.7/spark-warehouse)
(spark.repl.class.uri,spark://192.168.1.73:54812/classes)
(spark.jars,)
(spark.repl.class.outputDir,/private/var/folders/h1/9msx26wd1lj_hgcn5l5xpnxr0003g5/T/spark-e617f983-7c43-
53040/repl-25e6f76e-add5-47bd-bc5a-c19803ff64bf)
(spark.app.name,Spark shell)
(spark.executor.id,driver)
(spark.submit.deployMode,client)
(spark.master,local[*])
(spark.home,/Users/spark-user/spark-2.1.1-bin-hadoop2.7)
(spark.sql.catalogImplementation,hive)
(spark.app.id,local-1496548726845)
```

Figure 2-19. *Default configurations in Spark shell application*

To see the complete set of available objects you can access from `spark` variable, you can leverage the Spark shell code completion features.

```
scala> spark.<tab>
```

Figure 2-20 shows the result this command.

```
scala> spark.
!=                        catalog          ensuring       implicits         range          streams        wait
##                        close            eq             isInstanceOf      read           synchronized   →
+                         conf             equals         listenerManager   readStream     table
->                        createDataFrame  experimental   ne                sparkContext   time
==                        createDataset    formatted      newSession        sql            toString
asInstanceOf              emptyDataFrame   getClass       notify            sqlContext     udf
baseRelationToDataFrame   emptyDataset     hashCode       notifyAll         stop           version
```

Figure 2-20. *A complete list of variables that can be accessed from the spark variable*

Upcoming chapters have more examples of using `spark` to interact with underlying Spark functionalities.

Introduction to Collaborative Notebooks

Collaborative Notebooks is a commercial product offered by Databricks, the original creator of the open source project called Apache Spark. According to the product documentation, Collaborative Notebooks is designed for data engineers, data scientists, and data analysts to perform data analysis and build machine learning models that support multiple languages, built-in data visualization, and automatic data versioning. It also provides Spark on demand compute infrastructure and can execute jobs for production data pipelines on a specific schedule. It is built around Apache Spark and provides four main value propositions to customers around the world.

- Fully managed Spark clusters

- An interactive workspace for exploration and visualization

- A production pipeline scheduler

- A platform for powering your favorite Spark-based applications

The Collaborative Notebooks product has two versions, the full platform and the community edition. The commercial edition is a paid product that provides advanced features such as creating multiple clusters, user management, and job scheduling.

The community edition is free and ideal for developers, data scientists, data engineers and anyone who wants to learn Apache Spark or try Databricks.

The following section cover the basic features of Collaborative Notebooks community edition. It provides an easy and intuitive environment to learn Spark, perform data analysis or build Spark applications. This section is not intended to be a comprehensive guide. For that, you can refer to the Databricks user guide (`https://docs.databricks.com/user-guide/index.html`).

To use Collaborative Notebooks, you need to sign up for a free account on the community edition at `https://databricks.com/try-databricks`. This signup process is simple and quick; an account can be created in a matter of minutes. Once the necessary information is provided and submitted in the sign-up form, you shortly receive an email from Databricks to confirm your email, which looks something like Figure 2-21.

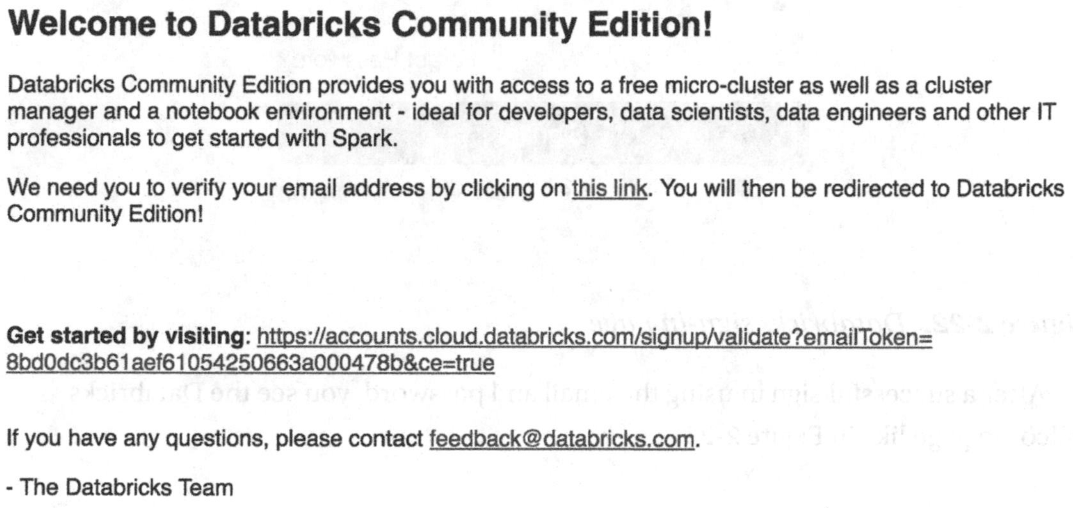

Welcome to Databricks Community Edition!

Databricks Community Edition provides you with access to a free micro-cluster as well as a cluster manager and a notebook environment - ideal for developers, data scientists, data engineers and other IT professionals to get started with Spark.

We need you to verify your email address by clicking on this link. You will then be redirected to Databricks Community Edition!

Get started by visiting: https://accounts.cloud.databricks.com/signup/validate?emailToken=8bd0dc3b61aef61054250663a000478b&ce=true

If you have any questions, please contact feedback@databricks.com.

- The Databricks Team

Figure 2-21. *Databricks email to confirm your email address*

Clicking the URL link shown in Figure 2-21 takes you to the Databricks sign-in form, as shown in Figure 2-22.

Figure 2-22. *Databricks sign-in page*

After a successful sign in using the email and password, you see the Databricks welcome page like in Figure 2-23.

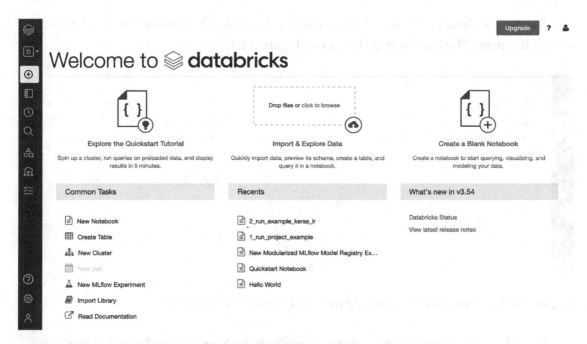

Figure 2-23. *Databricks welcome page*

Over time, the welcome page may evolve, so it does not look exactly like Figure 2-23. Feel free to explore the tutorial or the documentation.

This section aims to create a notebook in Databricks so that you can learn the commands covered in the previous section. The following are the main steps.

1. Create a cluster.

2. Create a folder.

3. Create a notebook.

Create a Cluster

One of the coolest features of the community edition (CE) is that it provides a single node Spark cluster with 15 GB of memory for free. At the time of writing this book, this single node cluster is hosted on the AWS cloud. Since the CE account is free, it provides the capability to create multiple clusters simultaneously. A cluster continues to stay up as long as it is being used. If it is idle for two hours, it automatically shuts down. This means you don't have to proactively shut down the cluster.

To create a cluster, click the Clusters icon in the vertical navigation bar on the left side of the page. The Cluster page looks like Figure 2-24.

Figure 2-24. *DataBricks Cluster page with no active clusters*

Now click the Create Cluster button to bring up the New Cluster form that looks like Figure 2-25.

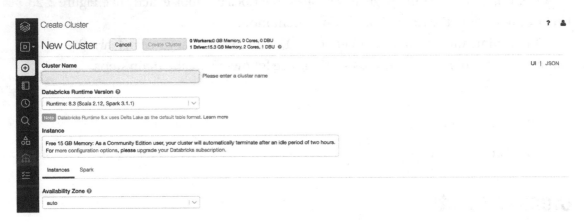

Figure 2-25. *Create Cluster form*

The only required field on this form is the cluster name. Table 2-4 describes each field.

Table 2-4. *Databricks New Cluster Form Fields*

Name	Description
Cluster Name	A unique name to identify your cluster. The name can have space between each word; for example, "my spark cluster".
Databricks Runtime Version	Databricks supports many versions of Spark. For learning purposes, select the latest version, which is automatically filled for you. Each version is tied to a specific AWS image.
Instance	For the CE edition, there isn't any other choice.
AWS – Availability Zone	This allows you to decide which AWS Availability Zone your single node cluster runs in. The options may look different based on your location.
Spark – Spark Config	This allows you to specify any application-specific configurations that should be used to launch the Spark cluster. Examples include JVM configurations to turn on certain Spark features.

Once you enter a cluster name, click the Create Cluster button. It can take up to 10 minutes to create a single node Spark cluster. If needed, try switching to a different availability zone if the default one takes a long time. Once a Spark cluster is successfully created, a green dot appears next to the cluster name, as shown in Figure 2-26.

Figure 2-26. *After a cluster is created successfully*

Feel free to explore by clicking the name of your cluster or the various links on this page. If you try to create another Spark cluster by following the same steps, it won't allow you to do so.

To terminate an active Spark cluster, click the square block under the Actions column.

For more information on creating and managing a Spark cluster in Databricks, go to `https://docs.databricks.com/user-guide/clusters/index.html`.

Let's move on to the next step, creating a folder.

Create a Folder

Before going into how to create a folder, it is worth it to take a moment to describe the workspace concept in Databricks. The easiest way to think about workspace is to treat it as the file system on your computer, which means one can leverage its hierarchical property to organize the various notebooks.

To create a folder, click the Workspace icon in the vertical navigation bar on the left side of the page. The Workspace column slides out, as shown in Figure 2-27.

Figure 2-27. *Workspace column*

Now click the downward arrow in the upper right of the Workspace column, and the popup menu shows up (see Figure 2-28).

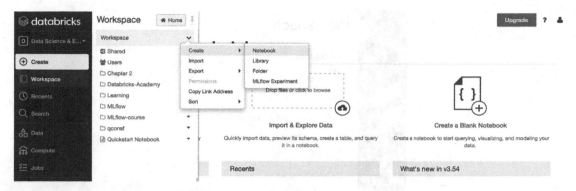

Figure 2-28. *Menu item for creating a folder*

Selecting the Create ➤ Folder menu item brings up the New Folder Name dialog box (see Figure 2-29).

Figure 2-29. *New Folder Name dialog box*

Now you can enter a folder name (i.e., Chapter 2), and click the Create Folder button to complete the process. The Chapter 2 folder should now appear in the Workspace column, as shown in Figure 2-30.

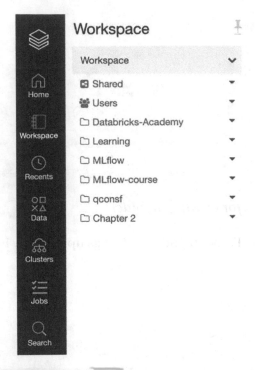

Figure 2-30. *Chapter 2 folder appears in the Workspace column*

Before creating a notebook, it is worth mentioning that there is an alternative way to create a folder. Place your mouse pointer anywhere in the Workspace column and right-click; the same menu options appear.

For more information on workspaces and creating folders, please go to https://docs.databricks.com/user-guide/workspace.html.

Create a Notebook

To create a Scala notebook in the Chapter 2 folder. First, select the Chapter 2 folder in the Workspace column. The Chapter 2 column slides out after the Workspace column, as shown in Figure 2-31.

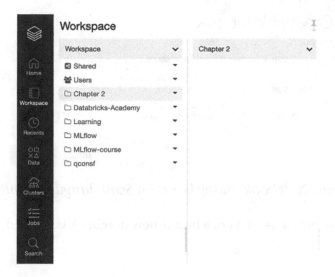

Figure 2-31. *Chapter 2 column appears to the right of Workspace column*

Now you can either click the downward arrow in the upper right of the Chapter 2 column or right-mouse click anywhere in the Chapter 2 column to bring the menu, as shown in Figure 2-32.

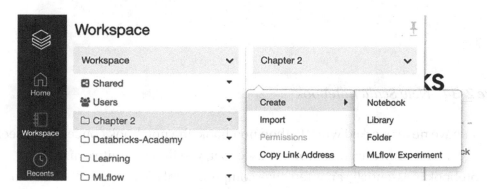

Figure 2-32. *Create notebook menu item*

Selecting the Notebook menu item brings up the Create Notebook dialog box. Give your notebook a name, and make sure to select the Scala option for the Language field. The value for the cluster should be automatically filled in because the CE edition can only have one cluster at a time. The dialog box should look something like Figure 2-33.

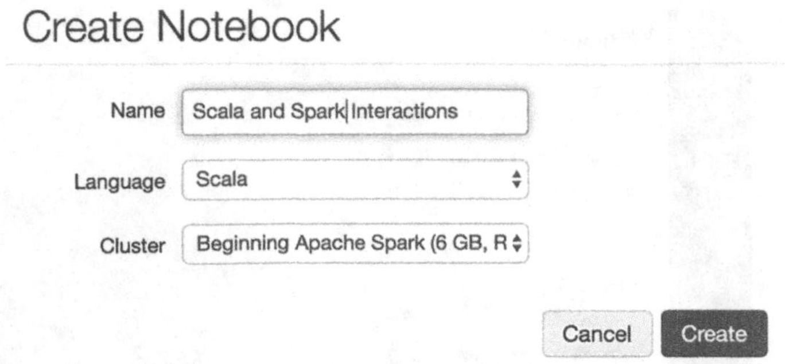

Create Notebook

Name Scala and Spark Interactions

Language Scala

Cluster Beginning Apache Spark (6 GB, R

Cancel   Create

Figure 2-33. *Create Notebook dialog box with Scala language option selected*

Once the Create button is clicked, a brand-new notebook is created, as shown in Figure 2-34.

Figure 2-34. *New Scala notebook*

If you have never worked with IPython notebook, the notebook concept may seem a bit strange at first. However, once you get used to it, you find it intuitive and fun.

A notebook is essentially an interactive computational environment (similar to Spark shell, but way better). You can execute Spark code, document your code with rich text using Markdown or HTML markup language and visualize the result of your data analysis with various types of chart and graph.

The following section covers only a few essential parts to help you be productive in using the Spark Notebook. For a comprehensive list of instructions on using and interacting with a Databricks notebook, please go to https://docs.databricks.com/user-guide/notebooks/index.html.

The Spark Notebook contains a collection of cells, and each one contains either a block of code to execute or markups for documentation purposes.

Note   A good practice of using the Spark Notebook is to break your data processing logic into multiple logical groups so each one resides in one or more cells. This is similar to the practice of developing maintainable software applications.

Let's divide the notebook into two sections. The first section contains the code snippets you entered in the "Basic Interactions with Scala" section. The second section contains the code snippets you entered in the "Basic Interactions with Spark" section.

Let's start with adding a Markdown statement to document the first section of your notebook by entering the following into the first cell (see Figure 2-35).

```
%md #### Basic Interactions with Scala
```

Figure 2-35. *Cell contains section header markup statement*

To execute that markup statement, make sure the mouse cursor is in cell 1, hold down the Shift key, and hit the Enter key. That is the shortcut for running code or markup statements in a cell. The result should look like Figure 2-36.

Basic Interactions with Scala

Figure 2-36. *The output of executing the markup statement*

Notice the Shift+Enter key combination executes the statements in that cell and creates a new cell below it. Now let's enter the "Hello World" example into the second cell and execute that cell. The output should look like Figure 2-37.

```
Cmd 1

Basic Interactions with Scala

Cmd 2

1  println("Hello from Spark Scala shell")

Hello from Spark Scala shell
Command took 0.10 seconds -- by hienluu@gmail.com at 6/4/2017, 10:01:19 AM on Beginning Apache Spark

Cmd 3

1                                                                    ▶▾ ⌄ — ✕
```

Figure 2-37. *The output of executing the println statement*

The remaining three code statements in the "Interactions with Scala" section are copied into the notebook (see Figure 2-38).

```
Cmd 3

1  val ages = Array(20, 50, 35, 41)
2  ages.foreach(println)

20
50
35
41
ages: Array[Int] = Array(20, 50, 35, 41)
Command took 0.81 seconds -- by hienluu@gmail.com at 6/4/2017, 10:04:57 AM on Beginning Apache Spark

Cmd 4

1  def isOddAge(age:Int) : Boolean = {
2    (age % 2) == 1
3  }

isOddAge: (age: Int)Boolean
Command took 0.13 seconds -- by hienluu@gmail.com at 6/4/2017, 10:05:41 AM on Beginning Apache Spark

Cmd 5

1  ages.filter(age => isOddAge(age)).foreach(println)

35
41
Command took 0.34 seconds -- by hienluu@gmail.com at 6/4/2017, 10:05:50 AM on Beginning Apache Spark
```

Figure 2-38. *The remaining code statements from the "Interactions with Scala" section*

Like Spark Scala shell, Scala Notebook is a full-blown Scala interactive environment where you can execute Scala code.

Now let's enter the second markup statement to denote the beginning of the second part of your notebook and the remaining code snippets in the "Interactions with Spark" section. Figure 2-39 shows the output.

```
%md #### Basic Interactions with Spark
```

Figure 2-39. *Output of the code snippets from Interactions with Spark section*

There are a few important notes to know when working with a Spark Notebook. It provides a very convenient auto-saving feature. The content of a notebook is automatically saved as you enter market statements or code snippets. In fact, the menu items under the File menu item don't have an option for saving a notebook.

Sometimes there is a need to create a new cell between two existing cells. One way to do this is to move the mouse cursor to the space between them, then click the plus icon that appears to create a new cell. Figure 2-40 shows what the plus icon looks like.

```
Cmd 8
  1  println("Spark version: " + spark.version)
Spark version: 2.1.1
Command took 0.09 seconds -- by hienluu@gmail.com at 6/4/2017, 10:34:48 AM on Beginning Apache Spark
Cmd 9                                              +
  1  spark.conf.getAll.foreach(println)
```

Figure 2-40. *Using plus icon to create a new cell between two existing cells*

Sometimes, you need to share your notebook with a co-worker who works in a remote office or other collaborators to either show off your awesome Spark knowledge or get their feedback on your data analysis. Simply click the File menu item at the top of your Spark notebook and select the Publish submenu item. Figure 2-41 shows what it looks like.

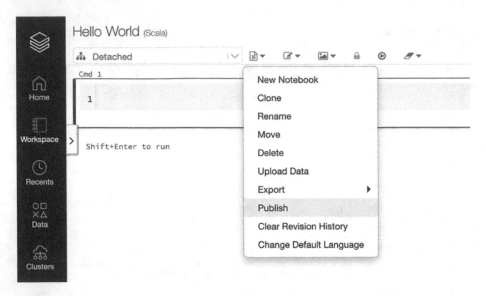

Figure 2-41. *Notebook publishing menu item*

Clicking the Publish submenu item brings up a confirmation dialog box (see Figure 2-42). If you follow through with it, the Notebook Published dialog box (see Figure 2-43) provides a URL that you can send to anyone in the world. With that URL, your co-worker or collaborators can view your notebook, or they can import it into their Databricks workspace.

Publish Notebook

Do you want to publish this notebook publicly? This action will overwrite any previously-published version of this notebook. Anyone with the link can view it and link will remain valid for 6 months.

☐ Don't show me this again

Cancel   Publish

Figure 2-42. *Publishing confirmation dialog box*

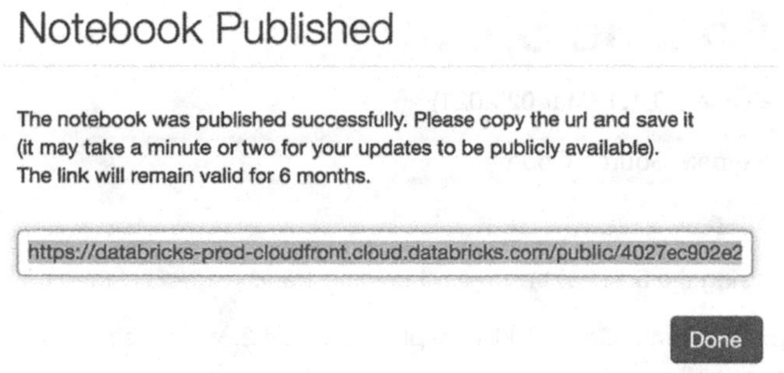

Notebook Published

The notebook was published successfully. Please copy the url and save it
(it may take a minute or two for your updates to be publicly available).
The link will remain valid for 6 months.

https://databricks-prod-cloudfront.cloud.databricks.com/public/4027ec902e2

Done

Figure 2-43. *Notebook published URL*

This section covers only the essential parts of using Databricks. Many other advanced features make it enticing to use Databricks as the platform for performing interactive data analysis or building advanced data solutions like machine learning models.

The CE provides a free account with a single node Spark cluster. Learning Spark through the Databricks product becomes so much easier than before. I highly recommend giving Databricks a try in your journey of learning Spark.

Setting up Spark Source Code

This section is geared toward software developers or anyone interested in learning how Spark works at the code level. Since Apache Spark is an open source project, its source code is publicly available to download from GitHub, examine and study how certain features were implemented. The Spark code is written in Scala by some of the smartest Scala programmers on the planet, so studying the Spark code is a great way to improve one's Scala programming skills and knowledge.

There are two ways to download Apache Spark source code to your computer. You can download it from the Spark download page at `http://spark.apache.org/downloads.html`, the same page used earlier to download the Spark binary file. This time, let's choose the Source Code package type, like in Figure 2-44.

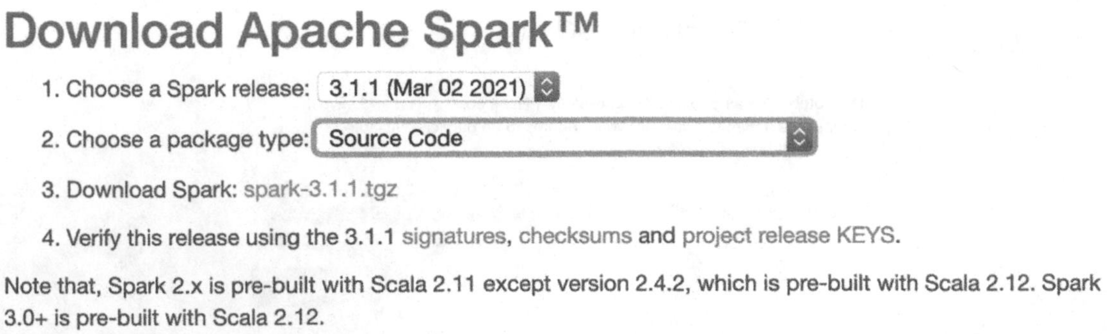

Download Apache Spark™

1. Choose a Spark release: 3.1.1 (Mar 02 2021)

2. Choose a package type: Source Code

3. Download Spark: spark-3.1.1.tgz

4. Verify this release using the 3.1.1 signatures, checksums and project release KEYS.

Note that, Spark 2.x is pre-built with Scala 2.11 except version 2.4.2, which is pre-built with Scala 2.12. Spark 3.0+ is pre-built with Scala 2.12.

Figure 2-44. *Apache Spark source download option*

To complete the source code download process, click the link on line #3 to download the compressed source code file. The final step is to uncompress the file into the directory your choice.

You can also use the `git clone` command to download Apache Spark source code from its GitHub repository. This requires an installation of git on your computer. Git is available for download at `https://git-scm.com/downloads`. The installation instructions are available at `https://git-scm.com/book/en/v2/Getting-Started-Installing-Git`. Once Git is properly installed on your computer, issue the following command to clone the Apache Spark git repository on GitHub (`https://github.com/apache/spark`).

```
git clone git://github.com/apache/spark.git
```

Once the Apache Spark source code is downloaded on your computer, go to `http://spark.apache.org/developer-tools.html` for information on how to import them into your favorite IDE.

Summary

- When it comes to learning Spark, there are a few options. You can either use the locally installed Spark or use the Collaborative Notebook Community Edition. These tools make it easy and convenient for anyone to learn Spark.

- The Spark shell is a powerful and interactive environment to learn Spark APIs and to analyze data interactively. There are two types of Spark shell, Spark Scala shell, and Spark Python shell.

- The Spark shell provides a set of commands to help its users to become productive.

- Collaborative Notebooks is a fully managed platform designed to simplify building and deploying data exploration, data pipelines, and machine learning solutions. The interactive workspace provides an intuitive way to organize and manage notebooks. Each notebook contains a combination of markup statements and code snippets. Sharing a notebook with others only requires a few mouse clicks.

- For software developers interested in learning about the internals of Spark, downloading and examining the Apache Spark source code is a great way to satisfy that curiosity.

CHAPTER 3

Spark SQL: Foundation

As Spark evolves and matures as a unified data processing engine with more features in each new release, its programming abstraction also evolves. The *resilient distributed dataset* (RDD) was the initial core programming abstraction when Spark was introduced to the world in 2012. In Spark version 1.6, a new programming abstraction called Structured APIs was introduced. This is the new and preferred way to handle data engineering tasks such as performing data processing or building data pipelines. The Structured APIs were designed to enhance developer productivity with easy-to-use, intuitive and expressive APIs. The new programming abstract requires the data available in a structured format, and the data computation logic needs to follow a certain structure. Armed with these two pieces of information, Spark can perform the necessary and sophisticated optimizations to speed up data processing applications.

Figure 3-1 shows how the Spark SQL component is built on top of the good old reliable Spark Core component. This layered architecture enables it to easily take advantage of any new improvements introduced in the Spark Core component.

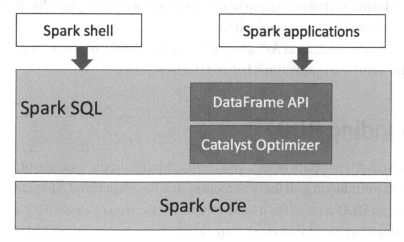

Figure 3-1. *Spark SQL components*

© Hien Luu 2021
H. Luu, *Beginning Apache Spark 3*, https://doi.org/10.1007/978-1-4842-7383-8_3

This chapter covers the Spark SQL module, which is designed for structured data processing. It provides an easy-to-use abstraction to express the data processing logic with the minimum amount of code, and underneath the cover, it intelligently performs necessary optimizations.

Spark SQL module consists of two main parts. The first one is the representations of the structure APIs called DataFrame and Dataset that define the high-level APIs for working with structured data. The DataFrame concept was inspired by the Python pandas DataFrame. The main difference is that a DataFrame in Spark can handle a large volume of data spread across many machines. The second part of the Spark SQL module is the Catalyst optimizer, which is responsible for all the complex machinery that works behind the scenes to make your life easier and ultimately speed up your data processing logic. One of the cool things that the Spark SQL module offers is executing SQL queries to perform data processing. With this capability, Spark can gain a new group of users called business analysts, who are very familiar with SQL language because it is one of the main tools they use regularly.

One main concept that differentiates structured data from unstructured data is the schema, which defines the data structure in the form of column names and associated data types. The schema concept is an integral part of Spark Structured APIs.

Structured data is often captured in a certain format. Some of the formats are text-based, and some of them are binary-based. Common formats for text data are CSV, XML, and JSON, and the common formats for binary data are Avro, Parquet, and ORC. Out of the box, the Spark SQL module makes it very easy to read data and write data from and to any of those formats. One unanticipated consequence of this versatility is that Spark can be used as a data format conversion tool.

Before going into Structured APIs, let's discuss the initial programming abstraction to better understand the motivations behind the new one.

Understanding RDD

To truly understand how Spark works, you must understand the essence of RDD. It provides a solid foundation and the abstraction that the Structured APIs are built upon. In short, an RDD represents a fault-tolerant collection of elements partitioned across the nodes of a cluster that can be operated in parallel. It consists of the following characteristics.

- A set of dependencies on parent RDDs

- A set of partitions, which are the chunks that make up the entire dataset

- A function for computing all the rows in the dataset

- The metadata about the partitioning scheme (optional)

- The location of where the data resides on the cluster (optional)

These five pieces of information are used by Spark runtime to schedule and execute the data processing logics expressed using the RDD operations.

The first three pieces of information make up the lineage information, which Spark uses for two purposes. The first is to determine the order of execution of RDDs and the second is for failure recovery.

The set of dependencies are essentially the input data to an RDD. This information is needed to reproduce the RDD in failure scenarios, and therefore it provides the resiliency characteristic.

The set of partitions enables Spark to execute the computation logic in parallel to speed up the computation time.

The last part that Spark needs to produce the RDD output is the compute function, which is provided by Spark users. The compute function is sent to each executor in the cluster to execute against each row in each partition.

The RDD abstraction is both simple and flexible. The flexibility has a drawback, where Spark has no insights into the user's intentions. It has no idea whether the computation logic is performing data filtering, joining, or aggregation. Therefore, Spark can't perform any optimizations, such as performing predicate pushdowns to reduce the amount of data to read from the input sources, recommending a more efficient join type to speed up the computation, or pruning the columns that are no longer needed by the output.

Introduction to the DataFrame API

A DataFrame is an immutable, distributed collection of data organized into rows. Each one consists of a set of columns and each column has a name and an associated type. In other words, this distributed collection of data has a structure defined by a schema. If you are familiar with the table concept in a relational database management system

(RDBMS), you realize that a DataFrame is essentially equivalent. A generic Row object represents each row in the DataFrame. Unlike RDD APIs, DataFrame APIs offer a set of domain specific operations that are relational and have rich semantics. You learn more about these APIs in upcoming sections. Like the RDD APIs, the DataFrame APIs are classified into two types: transformation and action. The evaluation semantics are identical in RDD. Transformations are lazily evaluated, and actions are eagerly evaluated.

A DataFrame can be created by reading data from many structured data sources and by reading data from tables in Hive or other databases. In addition, the Spark SQL module provides APIs to easily convert an RDD to a DataFrame by providing the schema information about the data in the RDD. The DataFrame API is available in Scala, Java, Python, and R.

Creating a DataFrame

There are many ways to create a DataFrame; one common thing among them is providing a schema, either implicitly or explicitly.

Creating a DataFrame from RDD

Let's start with creating a DataFrame from an RDD. Listing 3-1 first creates an RDD with two columns of integers. Then it calls the toDF implicit function that converts an RDD to a DataFrame using the specified column names. The column types are inferred from the data values in the RDD. Listing 3-2 shows two commonly used functions in a DataFrame, printSchema, and show. The printSchema function prints out the column names and their associated type to the console. The function prints out the data in a DataFrame in a tabular format. By default, it displays 20 rows. To change the default number of rows to display, you can pass a number to the show function. Listing 3-3 is an example of specifying the number of rows to display.

Listing 3-1. Creating DataFrame from an RDD of Numbers

```
import scala.util.Random
val rdd = spark.sparkContext.parallelize(1 to 10).map(x => (x, Random.
nextInt(100)* x))
val kvDF = rdd.toDF("key","value")
```

Listing 3-2. Print Schema and Show the Data of a DataFrame

```
kvDF.printSchema
|-- key: integer (nullable = false)
|-- value: integer (nullable = false)

kvDF.show
+----+-------+
| key|  value|
+----+-------+
|   1|     58|
|   2|     18|
|   3|    237|
|   4|     32|
|   5|     80|
|   6|    210|
|   7|    567|
|   8|    360|
|   9|    288|
|  10|    260|
+----+-------+
```

Listing 3-3. Call show Function to Display 5 Rows in Tabular Format

```
kvDF.show(5)
+----+------+
| key| value|
+----+------+
|   1|    59|
|   2|    60|
|   3|    66|
|   4|   280|
|   5|    40|
+----+------+
```

Note   The actual numbers in the value column may look different for you because they are generated randomly by calling the Random.nextInt() function.

Another way of creating a DataFrame is by specifying an RDD and a schema, which can be programmatically created. Listing 3-4 first creates an RDD using an array of Row objects, where each row object contains three columns. It creates a schema programmatically and finally provides the RDD and schema to the `createDataFrame` function to convert to a DataFrame. Listing 3-5 shows the schema and the data in the peopleDF DataFrame.

Listing 3-4. Create a DataFrame from a RDD with a Schema Created Programmatically

```
import org.apache.spark.sql.Row
import org.apache.spark.sql.types._

val peopleRDD = spark.sparkContext.parallelize(Array(Row(1L, "John
Doe",  30L),Row(2L, "Mary Jane", 25L)))

val schema = StructType(Array(
        StructField("id", LongType, true),
        StructField("name", StringType, true),
        StructField("age", LongType, true)
))

val peopleDF = spark.createDataFrame(peopleRDD, schema)
```

Listing 3-5. Display Schema of peopleDF and Its Data

```
peopleDF.printSchema
 |-- id: long (nullable = true)
 |-- name: string (nullable = true)
 |-- age: long (nullable = true)

peopleDF.show
+----+-------------+----+
| id |        name | age|
+----+-------------+----+
|   1|    John Doe|  30|
|   2|   Mary Jane|  25|
+----+-------------+----+
```

The ability to programmatically create a schema gives Spark applications the flexibility to adjust the schema based on some external configuration.

Each `StructField` object has three pieces of information: name, type, whether the value is nullable or not.

Each column type in a DataFrame is mapped to an internal Spark type, which can be a simple scalar type or a complex type. Table 3-1 references the available Scala type in Spark in the order of scalar type first and then the complex type last.

Table 3-1. *Spark Scala Type Reference*

Data Type	Scala Type
BooleanType	Boolean
ByteType	Byte
ShortType	Short
IntegerType	Int
LongType	Long
FloatType	Float
DoubleType	Double
DecimalType	java.math.BigDecial
StringType	String
BinaryType	Array[Byte]
TimestampType	java.sql.Timestamp
DateType	java.sql.Date
ArrayType	scala.collection.Seq
MapType	scala.collection.Map
StructType	org.apache.spark.sql.Row

Creating a DataFrame from a Range of Numbers

Spark 2.0 introduced a new entry point for Spark applications that primarily use DataFrame and Dataset APIs. This new entry point is represented by the `SparkSession` class, which has a convenient function called `range` that you can use to easily create a dataset with a single column with id as the name and LongType as the type. This function has a few variations that can take additional parameters to specify the end and the step. Listing 3-6 provides examples of using this function to create a DataFrame.

Listing 3-6. Examples Using SparkSession.range Function to Create a DataFrame

```
val df1 = spark.range(5).toDF("num").show
+-----+
|  num|
+-----+
|    0|
|    1|
|    2|
|    3|
|    4|
+-----+

spark.range(5,10).toDF("num").show
+-----+
|  num|
+-----+
|    5|
|    6|
|    7|
|    8|
|    9|
+-----+

spark.range(5,15,2).toDF("num").show
+------+
|   num|
+------+
|     5|
|     7|
|     9|
|    11|
|    13|
+------+
```

The last version of the range function takes three parameters. The first one represents the starting value, the second represents the end value (exclusive), and the last represents step size. Notice the range function can create only a single column DataFrame. Do you have any ideas about how to create a two-column DataFrame?

One option to create a multicolumn DataFrame uses Spark's implicits, which convert a collection of tuples inside a Scala Seq collection. Listing 3-7 is an example of Spark's toDF implicit.

Listing 3-7. Converting a Collection Tuples to a DataFrame Using Spark's toDF Implicit

```
val movies = Seq(("Damon, Matt", "The Bourne Ultimatum", 2007L),
                 ("Damon, Matt", "Good Will Hunting", 1997L))

val moviesDF = movies.toDF("actor", "title", "year")

moviesDF.printSchema
|-- actor: string (nullable = true)
|-- title: string (nullable = true)
|-- year: long (nullable = false)

moviesDF.show
+-----------+--------------------+------+
|      actor|               title| year|
+-----------+--------------------+------+
|Damon, Matt|The Bourne Ultimatum|  2007|
|Damon, Matt|   Good Will Hunting|  1997|
+-----------+--------------------+------+
```

These fun ways to create a DataFrame make it easy to learn and work with DataFrame APIs without loading the data from some external files. However, when you start performing serious data analysis with large datasets, it is imperative to know how to load data from external data sources, which is covered next.

Creating a DataFrame from Data Sources

Out of the box, Spark SQL supports a set of built-in data sources, where each one is mapped to a data format. The data source layer in the Spark SQL module is designed to be extensible, so custom data sources can be easily integrated into the DataFrame APIs. The Spark community writes hundreds of custom data sources, and it is not too difficult to implement them.

The two main classes in Spark for reading and writing data are `DataFrameReader` and `DataFrameWriter`, respectively. This section covers working with the APIs in the `DataFrameReader` class and the various available options when reading data from a specific data source.

An instance of the `DataFrameReader` class is as available as the `read` variable of the `SparkSession` class. You can refer to it from a Spark shell or in a Spark application, as shown in Listing 3-8.

Listing 3-8. Using read Variable from SparkSession

```
spark.read
```

The common pattern for interacting with DataFrameReader is described in Listing 3-9.

Listing 3-9. Common Pattern for Interacting with DataFrameReader

```
spark.read.format(...).option("key", value").schema(...).load()
```

Table 3-2 describes the three main pieces of information used when reading data: format, option, and schema. More on these three pieces of information is discussed in later in the chapter.

Table 3-2. *Main Information on DataFrameReader*

Name	Optional	Comments
format	No	It can be one of the built-in data sources or custom format. For a built-in format, you can use a short name (json, parquet, jdbc, orc, csv, text). For a custom data source, it requires providing a fully qualified name. See Listing 3-10 for examples.
option	Yes	DataFrameReader has a set of default options for each of the data source formats. You can override those default values by providing a value as the `option` function.
schema	Yes	Some data sources have the schema embedded in the data files, especially Parquet and ORC. In those cases, the schema is automatically inferred. For other cases, you may need to provide a schema.

Listing 3-10. Specifying Data Source Format

```
spark.read.json("<path>")
spark.read.format("json")

spark.read.parquet("<path>")
spark.read.format("parquet")

spark.read.jdbc
spark.read.format("jdbc")

spark.read.orc("<path>")
spark.read.format("orc")

spark.read.csv("<path>")
spark.read.format("csv")

spark.read.text("<path>")
spark.read.format("text")

// custom data source - fully qualified package name
spark.read.format("org.example.mysource")
```

Table 3-3 describes Spark's six built-in data sources and provides comments for each of them.

Table 3-3. *Spark's Built-in Data Sources*

Name	Data Format	Comments
Text file	Text	No structure.
CSV	Text	Comma-separated values. Can specify another delimiter. The column name can be referred from the header.
JSON	Text	Popular semistructured format. Column name and data type are inferred automatically
Parquet	Binary	(Default format) The popular binary format in the Hadoop community.
ORC	Binary	Another popular binary format in the Hadoop community.
JDBC	Binary	The common format for reading and writing to RDBMS.

Creating a DataFrame by Reading Text Files

Text files contain unstructured data. As it is read into Spark, each line becomes a row in the DataFrame. There are a lot of free books that are available for download in plain text format at www.gutenberg.org. For plain text files, one common way to parse the words is by splitting each line with a space delimiter. This is similar to how a typical word count example works. Listing 3-11 is an example of a README text file.

Listing 3-11. Read README.md File as a Text File from Spark Shell

```
val textFile = spark.read.text("README.md")

textFile.printSchema
|-- value: string (nullable = true)

// show 5 lines and don't truncate
textFile.show(5, false)
```

```
+----------------------------------------------------------------------+
|value                                                                 |
+----------------------------------------------------------------------+
|# Apache Spark                                                        |
|                                                                      |
|Spark is a fast and general cluster computing system for Big Data. It provides |
|high-level APIs in Scala, Java, Python, and R, and an optimized engine that    |
|supports general computation graphs for data analysis. It also supports a      |
+----------------------------------------------------------------------+
```

If a text file contains a delimiter that you can use to parse the columns in each line, then it is better to read it using CSV format, which is covered in the following section.

Creating a DataFrame by Reading CSV Files

One of the popular text file formats is CSV, which stands for *comma-separated values*. Popular tools like Microsoft Excel can easily import and export data in CSV format. The CSV parser in Spark is designed to be flexible such that it can parse a text file using a user-provided delimiter. The comma delimiter just happens to be the default one. This means you can use CSV format to read tab-separated value text files or other text files with an arbitrary delimiter.

Some CSV files have a header, and some don't. Since a column value may contain a comma, it is a common and good practice to escape it using a special character. Table 3-4 describes commonly used options when working with CSV format. For a complete list of options, please see the CSVOptions class at https://github.com/apache/spark.

Table 3-4. *CSV Common Options*

Key	Value(s)	Default	Description
sep	Single character	,	The single character value used as a delimiter for each column.
header	true,false	false	If the value is true, it means the first line in the file represents the column names.
escape	Any character	\	The character to use to escape the character in the column value is the same as sep.
inferSchema	true,false	false	Whether Spark should try to infer the column type based on column value.

Specifying the header and inferSchema options as true won't require you to specify the schema. Otherwise, you need to define a schema by hand or programmatically and pass it into the schema function. If the inferSchema option is false and no schema is provided, Spark assumes the data type for all the columns to be the string type.

The data file you are using as an example is called movies.csv in the data/chapter4 folder. This file contains a header for each column: actor, title, year. Listing 3-12 provides a few examples of reading CSV files.

Listing 3-12. Read CSV Files with Various Options

```
val movies = spark.read.option("header","true").csv("<path>/book/chapter4/
data/movies/movies.csv")

movies.printSchema
 |-- actor: string (nullable = true)
 |-- title: string (nullable = true)
 |-- year: string (nullable = true)

// now try to infer the schema
val movies2 = spark.read.option("header","true").
option("inferSchema","true")
                        .csv("<path>/book/chapter4/data/movies/movies.csv")
```

```
movies2.printSchema
 |-- actor: string (nullable = true)
 |-- title: string (nullable = true)
 |-- year: integer (nullable = true)

// now try to manually provide a schema
import org.apache.spark.sql.types._
val movieSchema = StructType(Array(StructField("actor_name", StringType, true),
                                   StructField("movie_title",
                                   StringType, true),
                                   StructField("produced_year",
                                   LongType, true)))
val movies3 = spark.read.option("header","true").schema(movieSchema)
                        .csv("<path>/book/chapter4/data/movies/
                        movies.csv")

movies3.printSchema
 |-- actor_name: string (nullable = true)
 |-- movie_title: string (nullable = true)
 |-- produced_year: long (nullable = true)

movies3.show(5)

+----------------+--------------+--------------+
|      actor_name|   movie_title| produced_year|
+----------------+--------------+--------------+
|McClure, Marc (I)|  Freaky Friday|          2003|
|McClure, Marc (I)|  Coach Carter|          2005|
|McClure, Marc (I)|    Superman II|          1980|
|McClure, Marc (I)|      Apollo 13|          1995|
|McClure, Marc (I)|      Superman|          1978|
+----------------+--------------+--------------+
```

The first example reads the file movies.csv with specifying the first line as the header. Spark can recognize column names. However, since the inferSchema option was not set to true, all the columns have string as the type. The second example added the inferSchema option, and Spark was able to identify column type. The third example provides a schema with column names different from what is in the header, so Spark uses the provided column names.

Now let's try to read in a text file with a different delimiter, not a comma. In this case, you specify a value for the sep option for Spark to use. Listing 3-13 shows a file called movies.tsv in the data/chapter4 folder.

Listing 3-13. Read a TSV File with CSV Format

```
val movies4 = spark.read.option("header","true").option("sep", "\t")
                                .schema(movieSchema).csv("<path>
                                /book/chapter4/data/movies/
                                movies.tsv")

movies.printSchema
|-- actor_name: string (nullable = true)
|-- movie_title: string (nullable = true)
|-- produced_year: long (nullable = true)
```

As you can see, it is quite easy to work with text files that have comma-separated values and other-separated values.

Creating a DataFrame by Reading JSON Files

JSON is a very well-known format in the JavaScript community. It is considered a semistructured format because each object (aka row) has a structure, and each column has a name. In the web application development space, JSON is widely used as a data format for transferring data between the backend server and the browser side. One of the strengths of JSON is that it provides a flexible format that can model any use case, and it can support nested structure. JSON has one disadvantage that is related to verbosity. The column name is repeated in each row in the data file (image your data file has 1 million rows).

Spark makes it easy to read data in a JSON file. However, there is one thing that you need to pay attention to. A JSON object can be expressed on a single line or across multiple lines, and this is something you need to let Spark know. Given that the JSON data file contains only column names and no data type, how can Spark come up with a schema? Spark tries its best to infer the schema by parsing a set of sample records. The number of records to sample is determined by the samplingRatio option, which has a default value of 1.0. Therefore, it is quite expensive to load a very large JSON file. In this case, you can lower the `samplingRatio` value to speed the data loading process. Table 3-5 describes a list of common options for the JSON format.

Table 3-5. *JSON Common Options*

Key	Value(s)	Default	Description
allowComments	true,false	false	Ignore comments in JSON file
multiLine	true,false	false	Treat the entire file as one large JSON object that spans across many lines
samplingRatio	0.3	1.0	The sampling size to read to infer the schema

Listing 3-14 shows two examples of reading JSON files. The first one simply reads a JSON file without overriding any option value. Notice Spark automatically detects the column name and data type based on the information in the JSON file. The second example specifies a schema.

Listing 3-14. Various Example of Reading a JSON File

```
val movies5 = spark.read.json("<path>/book/chapter4/data/movies/movies.json")

movies.printSchema
 |-- actor_name: string (nullable = true)
 |-- movie_title: string (nullable = true)
 |-- produced_year: long (nullable = true)

// specify a schema to override the Spark's inferring schema.
// producted_year is specified as integer type
import org.apache.spark.sql.types._
val movieSchema2 = StructType(Array(StructField("actor_name",
StringType, true),
                                    StructField("movie_title",
                                    StringType, true),
                                    StructField("produced_year",
                                    IntegerType, true)))
```

```
val movies6 = spark.read.option("inferSchema","true").schema(movieSchema2)
                                    .json("<path>/book/chapter4/data/
                                    movies/movies.json")

movies6.printSchema
 |-- actor_name: string (nullable = true)
 |-- movie_title: string (nullable = true)
 |-- produced_year: integer (nullable = true)
```

What happens when a column data type specified in the schema doesn't match the value in the JSON file? By default, when Spark encounters a corrupted record or runs into a parsing error, it set the value for all the columns in that row to be null. Instead of getting null values, you can tell Spark to fail fast. Listing 3-15 tells Spark's parsing logic to fail fast by specifying the mode option as failFast.

Listing 3-15. Parsing Error and How to Tell Spark to Fail Fast

```
// set data type for actor_name as BooleanType
import org.apache.spark.sql.types._
val badMovieSchema = StructType(Array(StructField("actor_name",
BooleanType, true),
                                    StructField("movie_title",
                                    StringType, true),
                                    StructField("produced_year",
                                    IntegerType, true)))

val movies7 = spark.read.schema(badMovieSchema)
                                    .json("<path>/book/chapter4/data/
                                    movies/movies.json")

movies7.printSchema
 |-- actor_name: boolean (nullable = true)
 |-- movie_title: string (nullable = true)
 |-- produced_year: integer (nullable = true)
```

```
movies7.show(5)
+----------+-----------+-------------+
|actor_name|movie_title|produced_year|
+----------+-----------+-------------+
|      null|       null|         null|
|      null|       null|         null|
|      null|       null|         null|
|      null|       null|         null|
|      null|       null|         null|
+----------+-----------+-------------+
```

```
// tell Spark to fail fast when facing a parsing error
val movies8 = spark.read.option("mode","failFast").schema(badMovieSchema)
                                 .json("<path>/book/chapter4/data/
                                 movies/movies.json")
```

```
movies8.printSchema
 |-- actor_name: boolean (nullable = true)
 |-- movie_title: string (nullable = true)
 |-- produced_year: integer (nullable = true)
```

```
// Spark will throw a RuntimeException when executing an action
movies8.show(5)
ERROR Executor: Exception in task 0.0 in stage 3.0 (TID 3)
java.lang.RuntimeException: Failed to parse a value for data type
BooleanType (current token: VALUE_STRING).
```

Creating a DataFrame by Reading Parquet Files

Parquet is one of the most popular open source columnar storage formats in the Hadoop ecosystem. It was created on Twitter. Its popularity is due to its self-describing data format, and it stores data in a highly compact structure by leveraging compressions. The columnar storage format is designed to work well with data analytics workload where only a small subset of columns are used during the data analysis. Parquet stores each column's data in a separate file; therefore, columns that are not needed in data analysis wouldn't have to be unnecessarily read in. It is quite flexible when it comes to supporting a complex data type with a nested structure. Text file formats like CSV and JSON are good for small files, and they are human-readable. Parquet is a much better file format for working with large datasets to reduce storage cost and speed up the reading step. If you peek at the `movies.parquet` file in the `chapter4/data/movies` folder, you see that its size is about one-sixth of the size of `movies.csv`.

Spark works extremely well with Parquet file format, and in fact, Parquet is the default file format for reading and writing data in Spark. Listing 3-16 shows an example of reading a Parquet file. Notice you don't need to provide a schema or ask Spark to infer the schema. Spark can retrieve the schema from the Parquet file.

A cool optimization that Spark does when reading data from Parquet is decompression and decoding in column batches, which considerably speeds up the reading.

Listing 3-16. Reading a Parquet File in Spark

```
// Parquet is the default format, so we don't need to specify the format
when reading
val movies9 = spark.read.load("<path>/book/chapter4/data/movies/movies.
parquet")
movies9.printSchema
 |-- actor_name: string (nullable = true)
 |-- movie_title: string (nullable = true)
 |-- produced_year: long (nullable = true)

// If we want to more explicit, we can specify the path to the parqet
function
val movies10 = spark.read.parquet("<path>/book/chapter4/data/movies/movies.
parquet")
movies10.printSchema
 |-- actor_name: string (nullable = true)
 |-- movie_title: string (nullable = true)
 |-- produced_year: long (nullable = true)
```

Creating a DataFrame by Reading ORC Files

Optimized Row Columnar (ORC) is another popular open source self-describing columnar storage format in the Hadoop ecosystem. It was created by Cloudera as a part of the initiative to massively speed up Hive. It is quite similar to Parquet in terms of efficiency and speed and was designed for analytics workload. Working with ORC files is just as easy as working with Parquet files. Listing 3-17 shows an example of creating a DataFrame from reading from an ORC file.

Listing 3-17. Reading ORC File in Spark

```
val movies11 = spark.read.orc("<path>/book/chapter4/data/movies/movies.orc")
movies11.printSchema
 |-- actor_name: string (nullable = true)
 |-- movie_title: string (nullable = true)
 |-- produced_year: long (nullable = true)

movies11.show(5)
+------------------------+-------------------+-------------+
|              actor_name|        movie_title| produced_year|
+------------------------+-------------------+-------------+
|         McClure, Marc (I)|       Coach Carter|         2005|
|         McClure, Marc (I)|       Superman II|         1980|
|         McClure, Marc (I)|          Apollo 13|         1995|
|         McClure, Marc (I)|           Superman|         1978|
|         McClure, Marc (I)| Back to the Future|         1985|
+------------------------+-------------------+-------------+
```

Creating a DataFrame from JDBC

JDBC is a standard application API for reading data from and writing data to a *relational database management system* (RDBMS). Spark has support for JDBC data source, which means you can use Spark to read data from and write data to any of the existing RDBMSs like MySQL, PostgreSQL, Oracle, SQLite, and so on. You need to provide a few important pieces of information when working with a JDBC data source: a JDBC driver for your RDBMS, a connection URL, authentication information, and a table name.

For Spark to connect to an RDBMS, it must have access to the JDBC driver JAR file at runtime. Therefore, you need to add the location of a JDBC driver to the Spark classpath. Listing 3-18 shows how to connect to MySQL from the Spark Shell.

Listing 3-18. Specifying a JDBC Driver When Starting the Spark Shell

```
./bin/spark-shell ../jdbc/mysql-connector-java-5.1.45/mysql-connector-
java-5.1.45-bin.jar  --jars ../jdbc/mysql-connector-java-5.1.45/mysql-
connector-java-5.1.45-bin.jar
```

Once the Spark shell successfully starts, you can quickly verify if Spark can connect to your RDBMS by using `java.sql.DriverManager`, as shown in Listing 3-19. This example is trying to test a connection to MySQL. The URL format is a bit different if your RDBMS is not MySQL, so consult the documentation of the JDBC driver you are using.

Listing 3-19. Testing Connection to MySQL in Spark Shell

```
import java.sql.DriverManager
val connectionURL = "jdbc:mysql://localhost:3306/<table>?user=<username>&pa
ssword=<password>"
val connection = DriverManager.getConnection(connectionURL)
connection.isClosed()
connection close()
```

If you didn't get any exception about the connection, the Spark shell could successfully connect to your RDBMS.

Table 3-6 describes the main options that you need to specify when using a JDBC driver. For a complete list of options, please consult `https://spark.apache.org/docs/latest/sql-programming-guide.html#jdbc-to-other-databases`.

Table 3-6. *Main Options for a JDBC Data Source*

Key	Description
url	The JDBC URL for Spark to connect to. At the minimum, it should contain the host, port and database name. For MySQL, it may look something like jdbc:mysql://localhost:3306/sakila.
dbtable	The name of the database table that Spark read data from or write data to.
driver	The class name of the JDBC driver that Spark instantiate to connect to the preceding URL. Consult the JDBC driver documentation that you are using. For MySQL Connector/J driver, the class name is `com.mysql.jdbc.Driver`.

Listing 3-20 shows an example of reading data from a film table of the Sakila database in a MySQL server.

Listing 3-20. Reading Data from a Table in MySQL Server

```
val mysqlURL= "jdbc:mysql://localhost:3306/sakila"
val filmDF = spark.read.format("jdbc").option("driver", "com.mysql.jdbc.
Driver")
                                            .option("url",
                                            mysqlURL)
                                            .option("dbtable",
                                            "film")
                                            .option("user",
                                            "<username>")
                                            .option("password",
                                            "<pasword>")
                                            .load()

filmDF.printSchema
 |-- film_id: integer (nullable = false)
 |-- title: string (nullable = false)
 |-- description: string (nullable = true)
 |-- release_year: date (nullable = true)
 |-- language_id: integer (nullable = false)
 |-- original_language_id: integer (nullable = true)
 |-- rental_duration: integer (nullable = false)
 |-- rental_rate: decimal(4,2) (nullable = false)
 |-- length: integer (nullable = true)
 |-- replacement_cost: decimal(5,2) (nullable = false)
 |-- rating: string (nullable = true)
 |-- special_features: string (nullable = true)
 |-- last_update: timestamp (nullable = false)

filmDF.select("film_id","title").show(5)
```

```
+-------+--------------------+
|film_id|               title|
+-------+--------------------+
|      1|    ACADEMY DINOSAUR|
|      2|      ACE GOLDFINGER|
|      3|     ADAPTATION HOLES|
|      4|     AFFAIR PREJUDICE|
|      5|         AFRICAN EGG|
+-------+--------------------+
```

When working with a JDBC data source, Spark pushes the filter conditions all the way down to the RDBMS as much as possible. By doing this, much of the data is filtered out at the RDBMS level, and therefore this speeds up the data filtering logic and dramatically reduces the amount of data Spark needs to read. This optimization is known as predicate pushdown, and Spark often does this when it knows the data source can support the filtering capability. Parquet is another data source that has this capability. The "Catalyst Optimizer" section in Chapter 4 provides an example of what this looks like.

Working with Structured Operations

Now that you know how to create a DataFrame, the next part is to learn how to manipulate or transform them using structured operations. Unlike the RDD operations, the structured operations are designed to be more relational, meaning the operations mirror the kind of expressions you can do with SQL, such as projection, filtering, transforming, joining, and so on. Similar to RDD operations, the structured operations are divided into two categories: transformation and action. The semantics of the structured transformations and actions are identical to the ones in RDDs. In other words, structured transformations are lazily evaluated, and structured actions are eagerly evaluated.

Structured operations are sometimes described as a domain-specific language (DSL) for distributed data manipulation. DSL is a computer language specialized for a particular application domain. In this case, the application domain is the distributed data manipulation. If you have ever worked with SQL, then it is easy to learn the structured operations.

Table 3-7 describes the commonly used DataFrame structured transformations. As a reminder, a DataFrame is immutable, and its transformation operation always returns a new DataFrame.

Table 3-7. *Commonly Used DataFrame Structured Transformations*

Operation	Description
select	Select one or more columns from an existing set of columns in the DataFrame. A more technical term for select is projection. During the projection process, columns can be transformed and manipulated.
selectExpr	Similar to select but provide powerful SQL expressions in transforming each column.
filter where	Both `filter` and where have the same semantics. where is more relational and similar to the where condition in SQL. They are both used for filtering rows based on the given boolean condition(s).
distinct dropDuplicates	Remove duplicate rows from the DataFrame
sort orderBy	Sort the DataFrame by the provided column(s)
limit	Return a new DataFrame by taking the first "n" rows.
union	Combine the rows from two DataFrame and return it as a new DataFrame.
withColumn	Use to add a column or replace an existing column in the DataFrame
withColumnRenamed	Renames an existing column. If a given column name doesn't exist in the schema, then it is a no-op.
drop	Drop one or more columns from DataFrame. The operation does nothing if schema doesn't contain the given column name(s)
sample	Randomly select a set of rows based on the given fraction, an optional seed value, and an optional replacement option.
randomSplit	Split the DataFrame into one or more DataFrames based on the given weights. Splits the master dataset into training and test datasets in the machine learning process.
join	Join two DataFrames. Spark supports many types of joins. More information is covered in the next chapter.
groupBy	Group the DataFrame by one or more columns. A common pattern is to perform aggregation after the groupBy. More information is covered in the next chapter.

Working with Columns

Most of the DataFrame structured operations in Table 3-7 require you to specify one or more columns. For some, the columns are specified in a string; for others, the columns need to be specified as instances of the Column class. It is completely fair to question why there are two options and when to use what. To answer those questions, you need to understand the functionality the Column class provides. At a high level, the Column class's functionality can be broken down into the following categories.

- Mathematical operations, like addition, multiplication, and so forth

- Logical comparisons between column value or a literal, such as equality, greater than, and less than

- String pattern matching, such as starting with, ending with, and so on.

For a complete list of available functions in the Column class, refer to the Scala documentation at https://spark.apache.org/docs/latest/api/scala/index. html#org.apache.spark.sql.Column.

With an understanding of the functionality that the Column class provides, you can conclude that whenever there is a need to specify a column expression, it is necessary to specify the column as an instance of Column class rather than a string. The upcoming examples make this clear.

There are different ways to refer to a column, which has created confusion in the Spark user community. A common question is when to use which one, and the answer is—it depends. Table 3-8 describes the available function options.

Table 3-8. *Ways to Refer to a Column*

Function	Example	Description
""	"columnName"	Refers to column as string type.
col	col("columnName")	The col function returns an instance of the Column class.
column	column("columnName")	Similar to col, this function returns an instance of the Column class.
$	$"columnName"	A syntactic sugar way of constructing a Column class in Scala only.
'(tick)	'columnName	A syntactic sugar way of constructing a Column class in Scala by leveraging Scala symbol literals feature.

Both col and column functions are synonymous, and both are available in Scala and Python Spark APIs. If you often switch between Spark Scala and Python APIs, then it makes sense to use the col function, so there is consistency in your code. If you mostly or exclusively use Spark Scala APIs, then my recommendation is to use ' (apostrophe symbol) because there is only a single character to type. The DataFrame class has its own col function, which disambiguates between columns with the same name from two or more DataFrames when performing a join. Listing 3-21 provides examples of different ways to refer to a column.

Listing 3-21. Different Ways of Referring to Columns

```
import org.apache.spark.sql.functions._

val kvDF = Seq((1,2),(2,3)).toDF("key","value")

// to display column names in a DataFrame, we can call the columns function
kvDF.columns
Array[String] = Array(key, value)

kvDF.select("key")
kvDF.select(col("key"))
kvDF.select(column("key"))
kvDF.select($"key")
kvDF.select('key)

// using the col function of DataFrame
kvDF.select(kvDF.col("key"))

kvDF.select('key, 'key > 1).show
+---+----------+
|key| (key > 1)|
+---+----------+
|  1|     false|
|  2|      true|
+---+----------+
```

This example illustrates a column expression, and therefore it is required to specify a column as an instance of the Column class. If the column was specified as a string, it results in a type mismatch error. More examples of column expressions are available in the examples of the various DataFrame structure operations.

77

Working with Structured Transformations

This section provides usage examples of the structured transformations listed in Table 3-7. To be consistent, all the examples consistently use a ' (apostrophe) to refer to column(s) in a DataFrame. To reduce redundancy, most of the examples refer to the movies DataFrame created by reading from a Parquet file (see Listing 3-22).

Listing 3-22. Creating the movies DataFame from a Parquet File

```
val movies = spark.read.parquet("<path>/chapter4/data/movies/movies.parquet")
```

select(columns)

This transformation commonly performs projection, selecting all or a subset of columns from a DataFrame. During the selection, each column can be transformed via a column expression. There are two variations of this transformation. One takes the column as a string, and the other takes columns as the Column class. This transformation doesn't permit you to mix the column type when using one of these two variations. Listing 3-23 is an example of the two variations.

Listing 3-23. Two Variations of Select Transformation

```
movies.select("movie_title","produced_year").show(5)
+--------------------+-------------+
|         movie_title| produced_year|
+--------------------+-------------+
|        Coach Carter|         2005|
|         Superman II|         1980|
|           Apollo 13|         1995|
|            Superman|         1978|
|    Back to the Future|         1985|
+--------------------+-------------+

// using a column expression to transform year to decade
movies.select('movie_title,('produced_year - ('produced_year % 10)).
as("produced_decade")).show(5)
```

```
+------------------------+----------------+
|            movie_title| produced_decade|
+------------------------+----------------+
|           Coach Carter|            2000|
|            Superman II|            1980|
|              Apollo 13|            1990|
|               Superman|            1970|
|      Back to the Future|           1980|
+------------------------+----------------+
```

The second example requires two column expressions: modulo and subtraction. Both are implemented by modulo (%) and subtraction (-) functions in the Column class (see the Scala documentation). By default, Spark uses the column expression as the name of the result column. To make it more readable, the as function is renames it to a more human-readable column name. As an astute reader, you can probably figure out the select transformation that can add one or more columns to a DataFrame.

selectExpr(expressions)

This transformation is a variant of the select transformation. The one big difference is that it accepts one or more SQL expressions rather than columns. However, both are essentially performing the same projection task. SQL expressions are powerful and flexible constructs that allow you to express column transformation logic naturally, just like the way you think. You can express SQL expressions in a string format, and Spark parses them into a logical tree to evaluate them in the right order.

If you want to create a new DataFrame with all the columns in the movies DataFrame and introduce a new column to represent the decade a movie was produced in, do something like what's shown in Listing 3-24.

Listing 3-24. Adding the Decade Column to Movies DataFrame using SQL
Expression

```
movies.selectExpr("*","(produced_year - (produced_year % 10)) as decade").
show(5)
+----------------+--------------------+-------------------+---------+
|      actor_name|         movie_title|      produced_year|   decade|
+----------------+--------------------+-------------------+---------+
|McClure, Marc (I)|        Coach Carter|               2005|     2000|
|McClure, Marc (I)|        Superman II |               1980|     1980|
|McClure, Marc (I)|           Apollo 13|               1995|     1990|
|McClure, Marc (I)|            Superman|               1978|     1970|
|McClure, Marc (I)|  Back to the Future|               1985|     1980|
+----------------+--------------------+-------------------+---------+
```

The combination of SQL expressions and built-in functions makes it easy to perform
a data analysis that otherwise take multiple steps. Listing 3-25 shows how easy it is to
determine the number of unique movie titles and unique actors in the movies dataset
in a single statement. The count function performs an aggregation over the entire
DataFrame.

Listing 3-25. Using SQL Expression and Built-in Functions

```
movies.selectExpr("count(distinct(movie_title)) as
movies","count(distinct(actor_name)) as actors").show
+---------+--------+
|   movies| actors |
+---------+--------+
|     1409|    6527 |
+---------+--------+
```

filler(condition), where(condition)

This transformation is straightforward. It filters out the rows that don't meet the given
condition, in other words, when the condition evaluates to false. A different way of
looking at the behavior of the filter transformation is that it returns only the rows that
meet the specified condition. The given condition can be simple or as complex as it
needs to be. Using this transformation requires knowing how to leverage a few logical

comparison functions in the Column class, like equality, less than, greater than, and inequality. Both the filter and where transformations have the same behavior, so pick the one that you are most comfortable with. The latter one is just a bit more relational than the former. Listing 3-26 shows a few examples of filtering.

Listing 3-26. Filter Rows with Logical Comparison Functions in Column Class

```
movies.filter('produced_year < 2000)
movies.where('produced_year > 2000)

movies.filter('produced_year >= 2000)
movies.where('produced_year >= 2000)

// equality comparison require 3 equal signs
movies.filter('produced_year === 2000).show(5)
+------------------+--------------------------+--------------+
|       actor_name|               movie_title| produced_year|
+------------------+--------------------------+--------------+
| Cooper, Chris (I)|         Me, Myself & Irene|          2000|
| Cooper, Chris (I)|                The Patriot|          2000|
|   Jolie, Angelina|        Gone in Sixty Sec...|          2000|
|    Yip, Françoise|            Romeo Must Die|          2000|
|    Danner, Blythe|           Meet the Parents|          2000|
+------------------+--------------------------+--------------+

// inequality comparison uses an interesting looking operator =!=
movies.select("movie_title","produced_year").filter('produced_year =!=
2000).show(5)
+--------------------+--------------+
|         movie_title| produced_year|
+--------------------+--------------+
|        Coach Carter|          2005|
|         Superman II|          1980|
|           Apollo 13|          1995|
|            Superman|          1978|
|  Back to the Future|          1985|
+--------------------+--------------+
```

```
// to combine one or more comparison expressions, we will use either the OR
and AND expression operator
movies.filter('produced_year >= 2000 && length('movie_title) < 5).show(5)
+----------------+------------+--------------+
|      actor_name| movie_title| produced_year|
+----------------+------------+--------------+
| Jolie, Angelina|        Salt|          2010|
|  Cueto, Esteban|         xXx|          2002|
|   Butters, Mike|         Saw|          2004|
|  Franko, Victor|          21|          2008|
|   Ogbonna, Chuk|        Salt|          2010|
+----------------+------------+--------------+

// the other way of accomplishing the result is by calling the filter
function two times
movies.filter('produced_year >= 2000).filter(length('movie_title) <
5).show(5)
```

distinct, dropDuplicates

These two transformations have identical behavior. However, dropDuplicates allows you to control which columns should be used in the deduplication logic. If none is specified, the deduplication logic uses all the columns in the DataFrame. Listing 3-27 shows different ways of counting how many movies are in the movies dataset.

Listing 3-27. Using distinct and dropDuplicates to Achieve the Same Goal

```
movies.select("movie_title").distinct.selectExpr("count(movie_title) as
movies").show
movies.dropDuplicates("movie_title").selectExpr("count(movie_title) as
movies").show

+--------+
|  movies|
+--------+
|    1409|
+--------+
```

In terms of performance, there is no difference between these two approaches because Spark transforms them into the same logical plan.

sort(columns), orderBy(columns)

Both transformations have the same semantics. The orderBy transformation is more relational than the other one. By default, the sorting is in ascending order, and it is easy to change it to descending. When specifying more than one column, it is possible to have a different order for each of those columns. Listing 3-28 has some examples.

Listing 3-28. Sorting the DataFrame in Ascending and Descending Order

```
val movieTitles = movies.dropDuplicates("movie_title")
                                   .selectExpr("movie_title",
"length(movie_title) as title_length", , "produced_year")

movieTitles.sort('title_length).show(5)
+-----------+-------------+--------------+
|movie_title| title_length| produced_year|
+-----------+-------------+--------------+
|         RV|            2|          2006|
|         12|            2|          2007|
|         Up|            2|          2009|
|         X2|            2|          2003|
|         21|            2|          2008|
+-----------+-------------+--------------+

// sorting in descending order
movieTitles.orderBy('title_length.desc).show(5)
+--------------------+-------------+--------------+
|         movie_title| title_length| produced_year|
+--------------------+-------------+--------------+
| Borat: Cultural L...|          83|          2006|
| The Chronicles of...|          62|          2005|
| Hannah Montana & ...|          57|          2008|
| The Chronicles of...|          56|          2010|
| Istoriya pro Rich...|          56|          1997|
+--------------------+-------------+--------------+
```

```
// sorting by two columns in different orders
movieTitles.orderBy('title_length.desc, 'produced_year).show(5)
+--------------------+------------+--------------+
|         movie_title| title_length| produced_year|
+--------------------+------------+--------------+
| Borat: Cultural L...|          83|          2006|
| The Chronicles of...|          62|          2005|
| Hannah Montana & ...|          57|          2008|
| Istoriya pro Rich...|          56|          1997|
| The Chronicles of...|          56|          2010|
+--------------------+------------+--------------+
```

Notice the title of the last two movies are at the same length, but their years are ordered in the correct ascending order.

limit(n)

This transformation returns a new DataFrame by taking the first *n* rows. This transformation is commonly used after the sorting is done to figure out the top *n* or bottom *n* rows based on the sorting order. Listing 3-20 shows an example of using the limit transformation to find the top ten actors with the longest names.

Listing 3-29. Using the limit Transformation to Figure Top Ten Actors with the Longest Name

```
// first create a DataFrame with their name and associated length
val actorNameDF = movies.select("actor_name").distinct.selectExpr("*",
"length(actor_name) as length")

// order names by length and retrieve the top 10
actorNameDF.orderBy('length.desc).limit(10).show
+--------------------------------+-------+
|          actor_name            | length|
+--------------------------------+-------+
|    Driscoll, Timothy 'TJ' James|     28|
|    Badalamenti II, Peter Donald|     28|
|    Shepard, Maridean Mansfield |     27|
|    Martino, Nicholas Alexander |     27|
```

```
|   Marshall-Fricker, Charlotte |     27|
|   Phillips, Christopher (III) |     27|
|   Pahlavi, Shah Mohammad Reza |     27|
|   Juan, The Bishop Don Magic  |     26|
|   Van de Kamp Buchanan, Ryan  |     26|
|     Lough Haggquist, Catherine |    26|
+-------------------------------+-------+
```

union(otherDataFrame)

You learned that a DataFrame is immutable. If there is a need to add more rows to an existing DataFrame, then the union transformation is useful for that purpose and combining rows from two DataFrames. This transformation requires that both DataFrames have the same schema, meaning both column names and their order must exactly match. Let's say one of the movies in the DataFrame is missing an actor, and you want to fix that issue. Listing 3-30 shows how to do that using union transformation.

Listing 3-30. Add a Missing Actor to the movies DataFrame

```
// the movie we want to add missing actor is "12"
val shortNameMovieDF = movies.where('movie_title === "12")
shortNameMovieDF.show

+--------------------+------------+---------------+
|         actor_name| movie_title| produced_year |
+--------------------+------------+---------------+
|     Efremov, Mikhail|         12|          2007|
|     Stoyanov, Yuriy|         12|          2007|
|      Gazarov, Sergey|         12|          2007|
| Verzhbitskiy, Viktor|         12|          2007|
+--------------------+------------+---------------+

// create a DataFrame with one row
import org.apache.spark.sql.Row
val forgottenActor = Seq(Row("Brychta, Edita", "12", 2007L))
val forgottenActorRDD = spark.sparkContext.parallelize(forgottenActor)
val forgottenActorDF = spark.createDataFrame(forgottenActorRDD,
shortNameMovieDF.schema)
```

```
// now adding the missing action
val completeShortNameMovieDF = shortNameMovieDF.union(forgottenActorDF)
completeShortNameMovieDF.union(forgottenActorDF).show
+---------------------+-----------+--------------+
|          actor_name| movie_title|  produced_year|
+---------------------+-----------+--------------+
|     Efremov, Mikhail|         12|          2007|
|      Stoyanov, Yuriy|         12|          2007|
|      Gazarov, Sergey|         12|          2007|
| Verzhbitskiy, Viktor|         12|          2007|
|        Brychta, Edita|         12|          2007|
+---------------------+-----------+--------------+
```

withColumn(colName, column)

This transformation adds a new column to a DataFrame. It requires two input parameters; a column name and a value in the form of a column expression. You can accomplish pretty much the same goal by using the selectExpr transformation. However, if the given column name matches one of the existing ones, that column is replaced with the given column expression. Listing 3-31 provides examples of adding a new column as well as replacing an existing one.

Listing 3-31. Add as Well Replacing a Column Using withColumn Transformation

```
// adding a new column based on a certain column expression
movies.withColumn("decade", ('produced_year - 'produced_year % 10)).show(5)
+------------------+-----------------------+--------------+----------+
|        actor_name|            movie_title| produced_year|    decade|
+------------------+-----------------------+--------------+----------+
| McClure, Marc (I)|           Coach Carter|          2005|      2000|
| McClure, Marc (I)|           Superman II|          1980|      1980|
| McClure, Marc (I)|              Apollo 13|          1995|      1990|
| McClure, Marc (I)|               Superman|          1978|      1970|
| McClure, Marc (I)|     Back to the Future|          1985|      1980|
+------------------+-----------------------+--------------+----------+
```

```
// now replace the produced_year with new values
movies.withColumn("produced_year", ('produced_year - 'produced_year % 10)).
show(5)
+-----------------+------------------+-------------+
|      actor_name|       movie_title| produced_year|
+-----------------+------------------+-------------+
| McClure, Marc (I)|      Coach Carter|         2000|
| McClure, Marc (I)|      Superman II|         1980|
| McClure, Marc (I)|        Apollo 13|         1990|
| McClure, Marc (I)|         Superman|         1970|
| McClure, Marc (I)| Back to the Future|        1980|
+-----------------+------------------+-------------+
```

withColumnRenamed(existingColName, newColName)

This transformation is strictly about renaming an existing column name in a DataFrame. It is fair to ask why in the world Spark provides this transformation. As it turns out, this transformation is useful in the following situations.

- To rename a cryptic column name to a more human friendly name. The cryptic column name can come from an existing schema that you don't control, such as when your company's partner produced the column you need in a Parquet file.

- Before joining two DataFrames that happen to have one or more same column name. This transformation can rename one or more columns in one of the two DataFrames, so you can refer to them easily after the join.

Notice that if the provided existingColName doesn't exist in the schema, Spark doesn't throw an error, and it silently does nothing. Listing 3-32 renames some of the column names in movies DataFrame to short names. By the way, this can be accomplished by using the select or selectExpr transformations as well. I leave that as an exercise for you.

Listing 3-32. Using withColumnRenamed Transformation to Rename Some of the Column Names

```
movies.withColumnRenamed("actor_name", "actor")
        .withColumnRenamed("movie_title", "title")
        .withColumnRenamed("produced_year", "year").show(5)
+------------------+--------------------+------+
|             actor|               title|  year|
+------------------+--------------------+------+
| McClure, Marc (I)|        Coach Carter|  2005|
| McClure, Marc (I)|        Superman II|  1980|
| McClure, Marc (I)|          Apollo 13|  1995|
| McClure, Marc (I)|           Superman|  1978|
| McClure, Marc (I)|  Back to the Future|  1985|
+------------------+--------------------+------+
```

drop(columnName1, columnName2)

This transformation simply drops the specified columns from the DataFrame. You can specify one or more column names to drop, but only the ones that exist in the schema are dropped, and the ones that don't are silently ignored. You can use the select transformation to drop columns by projecting out the columns you want to keep. However, if a DataFrame has 100 columns, and you want to drop a few, then this transformation is more convenient to use than the select transformation. Listing 3-33 provides examples of dropping columns.

Listing 3-33. Drop Two Columns, One Exists and the Other One Doesn't

```
movies.drop("actor_name", "me").printSchema
 |-- movie_title: string (nullable = true)
 |-- produced_year: long (nullable = true)
```

As you can see, the second column, "me", doesn't exist in the schema, and the drop transformation simply ignores it.

sample(fraction), sample(fraction, seed), sample(fraction, seed, withReplacement)

This transformation returns a randomly selected set of rows from the DataFrame. The number of the returned rows is approximately equal to the specified fraction, representing a percentage, and the value must be between 0 and 1. The seed seeds the random number generator, which generates a row number to include in the result. If a seed is not specified, then a randomly generated value is used. The `withReplacement` option determines whether a randomly selected row is placed back into the selection pool. In other words, when `withReplacement` is true, a particular selected row has the potential to be selected more than once. So, when would you need to use this transformation? It is useful when the original dataset is large and there is a need to reduce it down to a smaller size so you can quickly iterate on the data analysis logic. Listing 3-34 provides examples of using `sample` transformation.

Listing 3-34. Different ways of Using the sample Transformation

```
// sample with no replacement and a ratio
movies.sample(false, 0.0003).show(3)
+-------------------+--------------------+--------------+
|         actor_name|         movie_title| produced_year|
+-------------------+--------------------+--------------+
|    Lewis, Clea (I)|  Ice Age: The Melt...|        2006|
|    Lohan, Lindsay|   Herbie Fully Loaded|        2005|
|Tagawa, Cary-Hiro...|      Licence to Kill|        1989|
+-------------------+--------------------+--------------+

// sample with replacement, a ratio and a seed
movies.sample(true, 0.0003, 123456).show(3)
+--------------------+----------------+--------------+
|          actor_name|     movie_title| produced_year|
+--------------------+----------------+--------------+
| Panzarella, Russ (V)|   Public Enemies|        2009|
|         Reed, Tanoai|        Daredevil|        2003|
|        Moyo, Masasa|     Spider-Man 3|        2007|
+--------------------+----------------+--------------+
```

As you can see, the returned movies are pretty random.

randomSplit(weights)

This transformation is commonly used during the process of preparing the data to train machine learning models. Unlike the previous transformations, this one returns one or more DataFrames. The number of DataFrames it returns is based on the number of weights you specify. If the set of weights don't add up to 1, they are normalized accordingly to add up to 1. Listing 3-35 provides an example of splitting the movie DataFrame into three smaller ones.

Listing 3-35. Use randomSplit to Split movies DataFrame into Three Parts

```
// the weights need to be an Array
val smallerMovieDFs = movies.randomSplit(Array(0.6, 0.3, 0.1))

// let's see if the counts are added up to the count of movies DataFrame
movies.count
Long = 31393

smallerMovieDFs(0).count
Long = 18881

smallerMovieDFs(0).count + smallerMovieDFs(1).count + smallerMovieDFs(2).count
Long = 31393
```

Working with Missing or Bad Data

In reality, the data you often work with is not as clean as you would like. Maybe it's because the data evolves, and therefore some columns have values and some don't. It is important to deal with this kind of issue at the beginning of your data manipulation logic to prevent any unpleasant surprises that cause your long-running data processing job to stop working.

The Spark community recognizes the need to deal with missing data is a fact of life. Therefore, Spark provides a dedicated class called `DataFrameNaFunctions` to help in dealing with this inconvenient issue. An instance of `DataFrameNaFunctions` is available as the an member variable in the `DataFrame` class. There are three common ways of dealing with missing or bad data. The first way is to drop the rows that have missing values in one or more columns. The second way is to fill those missing values with user-provided values. The third way is to replace the bad data with something that you know how to deal with.

Let's start with dropping rows with missing data. You can tell Spark to drop rows where any column or only the specific columns have missing data. Listing 3-36 shows a few different ways of drop rows with missing data.

Listing 3-36. Dropping Rows with Missing Data

```
// first create a DataFrame with missing values in one or more columns
import org.apache.spark.sql.Row

val badMovies = Seq(Row(null, null, null),
                    Row(null, null, 2018L),
                    Row("John Doe", "Awesome Movie", null),
                    Row(null, "Awesome Movie", 2018L),
                    Row("Mary Jane", null, 2018L))
val badMoviesRDD = spark.sparkContext.parallelize(badMovies)
val badMoviesDF = spark.createDataFrame(badMoviesRDD, movies.schema)
badMoviesDF.show

+-----------+-----------------+--------------+
| actor_name|      movie_title| produced_year|
+-----------+-----------------+--------------+
|       null|             null|          null|
|       null|             null|          2018|
|   John Doe|    Awesome Movie|          null|
|       null|    Awesome Movie|          2018|
|  Mary Jane|             null|          2018|
+-----------+-----------------+--------------+

// dropping rows that have missing data in any column
// both of the lines below achieve the same output
badMoviesDF.na.drop().show
badMoviesDF.na.drop("any").show

+-----------+------------+--------------+
|actor_name| movie_title| produced_year|
+-----------+------------+--------------+
+-----------+------------+--------------+
```

```
// drop rows that have missing data in every single column
badMoviesDF.na.drop("all").show
+-----------+--------------+--------------+
| actor_name|   movie_title| produced_year|
+-----------+--------------+--------------+
|       null|          null|          2018|
|   John Doe| Awesome Movie|          null|
|       null| Awesome Movie|          2018|
|  Mary Jane|          null|          2018|
+-----------+--------------+--------------+

// drops rows that column actor_name has missing data
badMoviesDF.na.drop(Array("actor_name")).show
+-----------+--------------+--------------+
| actor_name|   movie_title| produced_year|
+-----------+--------------+--------------+
|   John Doe| Awesome Movie|          null|
|  Mary Jane|          null|          2018|
+-----------+--------------+--------------+
```

Working with Structured Actions

This section covers the structured actions. They have the same eager evaluation semantics as the RDD actions, so they trigger the computation of all the transformations that lead up to a particular action. Table 3-9 describes a list of structured actions.

Table 3-9. *Commonly Used Structured Actions*

Operation	Description
show() show(numRows) show(truncate) show(numRows, truncate)	Display the row in a tabular format. If numRows is not specified, it shows the top 20 rows. The truncate option controls whether to truncate a string column if it is longer than 20 characters.
head() first() head(n) take(n)	Return the first row. If n is specified, then it returns the first n rows. first is an alias for first. take(n) is an alias for first(n).
takeAsList(n)	Return the first n rows as a Java list. Be careful not to take too many rows; otherwise, it may cause an out-of-memory error on the application's driver process.
collect collectAsList	Return all the rows as an array or a Java list. Apply the same caution as the one described in takeAsList action.
count	Return the number of rows in the DataFrame.
describe	Compute common statistics about numeric and string columns in the DataFrame. Available statistics are count, mean, stddev, min, max, and arbitrary approximate percentiles.

Most of these are self-explanatory. The show action has been used in many examples in the structured transformation section.

Another interesting action is called describe, which is discussed next.

describe(columnNames)

Sometimes it is useful to have a general sense of the basic statistics of the data you are working with. This action can compute the basic statistics of string and numeric columns, such as count, mean, standard deviation, minimum, and maximum. You have the option to choose which string or numeric column(s) to compute the statistics for. Listing 3-37 is an example.

Listing 3-37. Use describe Action to Show the Statistics of produced_year Column

```
movies.describe("produced_year").show
+-----------+-------------------------+
|    summary|            produced_year|
+-----------+-------------------------+
|      count|                    31392|
|       mean|        2002.7964449541284|
|     stddev|         6.377236851493877|
|        min|                     1961|
|        max|                     2012|
+-----------+-------------------------+
```

Introduction to Datasets

At one point, there was a lot of confusion about the differences between the DataFrame and Dataset APIs. Given these options, it is fair to ask what the differences are between them, the advantages and disadvantages of each option, and when to use which one. Recognizing this huge confusion in the Spark user community, Spark designers decided to unify the DataFrame APIs with Dataset APIs in Spark 2.0 version to have one fewer abstraction for users to learn and remember.

Starting with the Spark 2.0 release, there is only one high-level abstraction called Dataset, which has two flavors: a strongly-typed API and an untyped API. The term DataFrame doesn't go away; instead, it has been redefined as an alias for a collection of generic objects in Dataset. From the code perspective, a DataFrame is essentially a type alias for Dataset[Row], where a Row is a generic untyped JVM object. A Dataset is a collection of strongly-typed JVM objects, represented by either a case class in Scala or a class in Java. Table 3-10 describes the Dataset API flavors available in each of the programming languages that Spark supports.

Table 3-10. *Dataset Flavors*

Language	Flavor
Scala	Dataset[T] and DataFrame
Java	Dataset[T]
Python	DataFrame
R	DataFrame

The Python and R languages have no compile-time type-safety; therefore, only the untyped Dataset APIs (a.k.a. DataFrame) are supported.

Consider the Dataset as a younger brother of DataFrame. Its unique properties include type safety and object-oriented. A Dataset is a strongly typed, immutable collection of data. Like a DataFrame, the data is mapped to a defined schema. However, there are a few important differences between a DataFrame and a Dataset.

- Each row in a Dataset is represented by a user-defined object so that you can refer to an individual column as a member variable of that object. This provides you with compile-type safety.

- The Dataset has helpers called encoders, which are smart and efficient encoding utilities that convert data in each user-defined object into a compact binary format. This translates into a reduction of memory usage when a Dataset is cached in memory and a reduction in the number of bytes when Spark needs to transfer over a network during the shuffling process.

In terms of limitations, the Dataset APIs are available in only strongly typed languages such as Scala and Java. There is the conversion cost associated with converting a Row object into a domain-specific object, and this cost can be a factor when a Dataset has millions of rows. At this point, a question should pop into your mind regarding when to use DataFrame APIs and Dataset APIs. The Dataset APIs are good for production jobs that need to run regularly and are written and maintained by a team of Data Engineers. For most interactive and explorative analysis use cases, using the DataFrame APIs is sufficient.

Note   A case class in the Scala language is like a JavaBean class in Java language; however, it has a few built-in interesting properties. An instance of a case class is immutable, and therefore it is commonly used to model domain-specific objects. In addition, it is easy to reason about the internal states of the instances of a case class because they are immutable. The toString and equals methods are automatically generated to make it easier to print out the case class's content and compare between case class instances. Scala case classes work well with the Scala pattern matching feature.

Creating Datasets

Before creating a Dataset, you need to define a domain-specific object to represent each row. There are a few ways to create a Dataset. The first way is to transform a DataFrame to a Dataset using the as(Symbol) function of the DataFrame class. The second way is to use the SparkSession.createDataset() function to create a Dataset from a collection of objects. The third way is to use the toDS implicit conversion utility. Listing 3-38 provides different examples of creating Datasets.

Listing 3-38. Different Ways of Creating Datasets

```
// define Movie case class
case class Movie(actor_name:String, movie_title:String, produced_year:Long)

// convert DataFrame to strongly typed Dataset
val moviesDS = movies.as[Movie]

// create a Dataset using SparkSession.createDataset() and the toDS
implicit function
val localMovies = Seq(Movie("John Doe", "Awesome Movie", 2018L),
                                    Movie("Mary Jane", "Awesome Movie",
                                    2018L))

val localMoviesDS1 = spark.createDataset(localMovies)
val localMoviesDS2 = localMovies.toDS()
localMoviesDS1.show
```

```
+------------+---------------+-------------+
|  actor_name|    movie_title|produced_year|
+------------+---------------+-------------+
|    John Doe|  Awesome Movie|         2018|
|  Mary Jane|  Awesome Movie|         2018|
+------------+---------------+-------------+
```

Among the different ways of creating Datasets, the first way is the most popular one. While transforming a DataFrame to a Dataset using a Scala case class, Spark performs a validation to ensure the member variable names in the Scala case class match up with column names in the schema of the DataFrame. If there is a mismatch, Spark lets you know.

Working with Datasets

Now that you have a Dataset, you can manipulate it using the transformations and actions. Earlier in the chapter, the columns in the DataFrame used one of these options. With a Dataset, each row is represented in a strongly typed object; therefore, you can just refer to the columns using the member variable names, which give you type safety and compile-time validation. If there is a misspelling in the name, the compiler flags them immediately during the development phase. Listing 3-39 are examples of manipulating a Dataset.

Listing 3-39. Manipulating a Dataset in a Type-Safe Manner

```
// filter movies that were produced in 2010 using
moviesDS.filter(movie => movie.produced_year == 2010).show(5)
+--------------------+--------------------+-------------+
|          actor_name|         movie_title|produced_year|
+--------------------+--------------------+-------------+
|    Cooper, Chris (I)|            The Town|         2010|
|     Jolie, Angelina|                Salt|         2010|
|     Jolie, Angelina|         The Tourist|         2010|
|      Danner, Blythe|       Little Fockers|         2010|
|  Byrne, Michael (I)|  Harry Potter and ...|         2010|
+--------------------+--------------------+-------------+
```

```
// displaying the title of the first movie in the moviesDS
moviesDS.first.movie_title
String = Coach Carter

// try with misspelling the movie_title and get compilation error
moviesDS.first.movie_tile
error: value movie_tile is not a member of Movie

// perform projection using map transformation
val titleYearDS = moviesDS.map(m => ( m.movie_title, m.produced_year))
titleYearDS.printSchema
 |-- _1: string (nullable = true)
 |-- _2: long (nullable = false)
```

```
// demonstrating a type-safe transformation that fails at compile time,
performing subtraction on a column with string type
```

```
// a problem is not detected for DataFrame until runtime
movies.select('movie_title - 'movie_title)
// a problem is detected at compile time
moviesDS.map(m => m.movie_title - m.movie_title)
error: value - is not a member of String
```

```
// take action returns rows as Movie objects to the driver
moviesDS.take(5)
Array[Movie] = Array(Movie(McClure, Marc (I),Coach Carter,2005),
Movie(McClure, Marc (I),Superman II,1980), Movie(McClure, Marc (I),Apollo
13,1995))
```

For those who use the Scala programming language regularly, working with Dataset strongly-typed APIs feels natural and gives you the impression that those objects in the Dataset reside locally.

When you use the Dataset strongly-typed APIs, Spark implicitly converts each Row instance to the domain-specific object that you provide. This conversion has some cost in terms of performance; however, it provides more flexibility.

One general guideline to help decide when to use a Dataset over DataFrame is the desire to have a higher degree of type-safety at compile time, which are important for complex ETL Spark jobs developed and maintained by multiple Data Engineers.

Using SQL in Spark SQL

In the big data era, SQL has been described as the lingua franca for big data analysis. One of the coolest features in Spark is the ability to use SQL to perform distributed data manipulation at scale. Data analysts who are proficient at SQL can now use Spark to perform data analysis on large datasets. One important note to remember is that SQL in Spark is designed for online analytical processing (OLAP) use cases, not online transaction processing use cases (OLTP). In other words, it is not applicable for low-latency use cases.

SQL has evolved and improved over time. Spark implements a subset of ANSI SQL:2003 revision, which most popular RDBMS servers support. Being compliant with this revision means Spark SQL data processing engine can be benchmarked using a widely used industry-standard decision support benchmark called TPC-DS.

In late 2016, Facebook began migrating some of its largest Hive workloads to Spark to take advantage of the power of the Spark SQL engine (see `https://code.facebook.com/posts/1671373793181703/apache-spark-scale-a-60-tb-production-use-case/`).

Note   Structure Query Language (SQL) is a domain-specific language that performs data analysis and manipulation of structured data organized in a table format. The concepts in SQL are based on relational algebra; however, it is an easy language to learn. One key difference between SQL and other programming languages like Scala or Python is that SQL is a declarative programming language, which means you express what you want to do with the data and let the SQL execution engine figure out how to perform the data manipulations as well as the necessary optimizations to speed up execution time. If you are new to SQL, there is a free course at this site at `www.datacamp.com/courses/intro-to-sql-for-data-science`.

Running SQL in Spark

Spark provides a few different options for running SQL in Spark.

- Spark SQL CLI (./bin/spark-sql)

- JDBC/ODBC server

- Programmatically in Spark applications

The first two options integrate Apache Hive to leverage its megastore, a repository that contains the metadata and schema information about the various systems and user-defined tables. This section covers only the last option.

A DataFrame and a Dataset are essentially like tables in a database. Before you can issue SQL queries to manipulate them, you need to register them as temporary views. Each view has a name, which is used as the table name in the select clause. Spark provides two levels of scoping for views. One is at the Spark session level. When a DataFrame is registered at this level, only the queries issued in the same session can refer to that DataFrame. The session-scoped level disappears when the associated Spark session is closed. The second scoping level is global, which means these views are available to SQL statements in all Spark sessions. All the registered views are maintained in the Spark metadata catalog that can be accessed via SparkSession. Listing 3-40 is an example of registering views and using the Spark catalog to inspect the metadata of the views.

Listing 3-40. Register the movies DataFrame as a Temporary View and Inspecting Metadata Catalog

```
// display tables in the catalog, expecting an empty list
spark.catalog.listTables.show
+--------+------------+---------------+------------+------------+
|   name|   database|   description|   tableType| isTemporary|
+--------+------------+---------------+------------+------------+
+--------+------------+---------------+------------+------------+

// now register movies DataFrame as a temporary view
movies.createOrReplaceTempView("movies")

// should see the movies view in the catalog
spark.catalog.listTables.show
+--------+----------+------------+----------+--------------+
|   name| database| description| tableType|   isTemporary|
+--------+----------+------------+----------+--------------+
| movies|     null|        null| TEMPORARY|          true|
+--------+----------+------------+----------+--------------+
```

```
// show the list of columns of movies view in catalog
spark.catalog.listColumns("movies").show
+-------------+------------+---------+---------+------------+-----------+
|         name| description| dataType| nullable| isPartition|   isBucket|
+-------------+------------+---------+---------+------------+-----------+
|   actor_name|        null|   string|     true|       false|      false|
|  movie_title|        null|   string|     true|       false|      false|
| produced_year|       null|   bigint|     true|       false|      false|
+-------------+------------+---------+---------+------------+-----------+

// register movies as global temporary view called movies_g
movies.createOrReplaceGlobalTempView("movies_g")
```

Listing 3-40 gives you a couple of views to select from. The programmatic way of issuing SQL queries is to use the `sql` function of `SparkSession` class. In the SQL statement, you have access to all SQL expressions and built-in functions. The `SparkSession.sql` function executes the given SQL query; it returns a DataFrame. The ability to issue SQL statements and use DataFrame transformations and actions provides you a lot of flexibility in how you choose to perform distributed data processing in Spark.

Listing 3-41 provides examples of issuing simple and complex SQL statements.

Listing 3-41. Executing SQL Statements in Spark

```
// simple example of executing a SQL statement without a registered view
val infoDF = spark.sql("select current_date() as today , 1 + 100 as value")
infoDF.show
+----------+--------+
|     today|   value|
+----------+--------+
|2017-12-27|     101|
+----------+--------+

// select from a view
```

```
spark.sql("select * from movies where actor_name like '%Jolie%' and
produced_year > 2009").show
+---------------+----------------+--------------+
|     actor_name|     movie_title| produced_year|
+---------------+----------------+--------------+
|Jolie, Angelina|            Salt|          2010|
|Jolie, Angelina| Kung Fu Panda 2|          2011|
|Jolie, Angelina|      The Tourist|          2010|
+---------------+----------------+--------------+

// mixing SQL statement and DataFrame transformation
spark.sql("select actor_name, count(*) as count from movies group by actor_
name")
        .where('count > 30)
        .orderBy('count.desc)
        .show
+--------------------+--------+
|          actor_name|   count|
+--------------------+--------+
|     Tatasciore, Fred|     38|
|        Welker, Frank|     38|
|   Jackson, Samuel L.|     32|
|        Harnell, Jess|     31|
+--------------------+--------+

// using a subquery to figure out the number movies produced each year.
// leverage """ to format multi-line SQL statement

spark.sql("""select produced_year, count(*) as count
                from (select distinct movie_title, produced_year from
                movies)
                group by produced_year""")
        .orderBy('count.desc).show(5)
```

```
+-------------------+--------+
|     produced_year| count|
+-------------------+--------+
|              2006|      86|
|              2004|      86|
|              2011|      86|
|              2005|      85|
|              2008|      82|
+-------------------+--------+
```

```
// select from a global view requires prefixing the view name with key word
'global_temp'
spark.sql("select count(*) from global_temp.movies_g").show
+--------+
|   count|
+--------+
|   31393|
+--------+
```

Instead of reading the data file through DataFrameReader class and registering the newly created DataFrame as a temporary view, there is a short and convenient way to issue SQL queries against a data file. Listing 3-42 is an example.

Listing 3-42. Issue SQL Query Against a Data File

```
spark.sql("SELECT * FROM parquet.`<path>/chapter4/data/movies/movies.
parquet`").show(5)
```

Writing Data Out to Storage Systems

At this point, you know how to read data from various file formats or a database server using DataFrameReader, and you know how to use SQL or transformations and actions of structured APIs to manipulate the data. At some point, you need to write the result of the data processing logic in the DataFrame to an external storage system (i.e., a local file system, HDFS, or Amazon S3). In a typical ETL data processing job, the results most likely be written out to some persistent storage system.

In Spark SQL, the `DataFrameWriter` class is responsible for the logic and complexity of writing out the data in a DataFrame to an external storage system. An instance of `DataFrameWriter` class is available to you as the `write` variable in the DataFrame class. The pattern for interacting with `DataFrameWriter` is similar to the interacting pattern of `DataFrameReader`. You can refer to it from a Spark shell or in a Spark application, as shown in Listing 3-43.

Listing 3-43. Using write Variable from DataFrame Class

```
movies.write
```

Listing 3-44 describes the common pattern for interacting with `DataFrameWriter`.

Listing 3-44. Common Interacting Pattern with DataFrameWriter

```
movies.write.format(...).mode(...).option(...).partitionBy(...).bucketBy(..
.).sortBy(...).save(path)
```

Similar to `DataFrameReader`, the default format is Parquet; therefore, it is unnecessary to specify a format if the desired output format is Parquet. The `partitionBy`, `bucketBy,` and `sortBy` functions control the directory structure of the output files in the file-based data sources. Structuring the directory layout based on reading patterns dramatically reduces the amount of data that needs to be read for analysis. You learn more about this later in the chapter. The input to the `save` function is a directory name, not a file name.

One of the important options in the `DataFrameWriter class is the save mode,` which controls how Spark handles the situation when the specified output location exists. Table 3-11 lists the various supported save modes.

Table 3-11. *Save Modes*

Mode	Description
append	This appends the DataFrame data to the list of files that already exist at the specified destination location.
overwrite	This completely overwrites any data files that already exist at the specified destination location with the data in the DataFrame.
error errorIfExists default	This is the default mode. If the specified destination location exists, then DataFrameWriter throws an error.
ignore	If the specified destination location exists, then simply do nothing. In other words, silently don't write out the data in the DataFrame.

Listing 3-45 shows a few examples of using the various combination of formats and modes

Listing 3-45. Using DataFrameWriter to Write Out Data to File-based Sources

```
// write data out in CVS format, but using a '#' as delimiter
movies.write.format("csv").option("sep", "#").save("/tmp/output/csv")

// write data out using overwrite save mode
movies.write.format("csv").mode("overwrite").option("sep", "#").save("/tmp/
output/csv")
```

The number of files written out to the output directory is corresponding to the number of partitions your DataFrame has. Listing 3-46 shows how to find out the number of partitions a DataFrame has.

Listing 3-46. Display the Number of DataFrame Partitions

```
movies.rdd.getNumPartitions
Int = 1
```

When the number of rows in a DataFrame is not large, there is a need to have a single output file to make it easier to share. A small trick to achieve this goal is to reduce the number of partitions in your DataFrame to one and then write it out. Listing 3-47 shows an example of how to do that.

Listing 3-47. Reduce the Number of Partitions in a DataFrame to 1

```
val singlePartitionDF = movies.coalesce(1)
```

The idea of writing data out using partitioning and bucketing is borrowed from the Apache Hive user community. As a rule of thumb, the partition by column should have low cardinality. In the `movies` DataFrame, the `produced_year` column is a good candidate for the partition by column. Let's say you want to write out the `movies` DataFrame with partitioning by the `produced_year` column. The DataFrameWriter writes out all the movies with the same `produced_year` into a single directory. The number of directories in the output folder corresponds to the number of years in the `movies` DataFrame. Listing 3-48 is an example of using `partitionBy` function.

Listing 3-48. Write the movies DataFrame Using Partition By produced_year Column

```
movies.write.partitionBy("produced_year").save("/tmp/output/movies ")

// the /tmp/output/movies directory will contain the following
subdirectories
produced_year=1961 to produced_year=2012
```

The directory names generated by the `partitionBy` option seems strange because each directory name consists of the partitioning column name and the associated value. These two pieces of information are used at the data reading time to choose which directory to read based on the data access pattern, and therefore it ends up reading much less data than otherwise.

The Trio: DataFrame, Dataset, and SQL

Now you know there are three different ways of manipulating structured data in the Spark SQL module. Table 3-12 shows where each option falls in the syntax and analysis spectrum.

Table 3-12. *Syntax and Analysis Errors Spectrum*

	SQL	DataFrame	Dataset
System Errors	Runtime	Compile time	Compile time
Analysis Errors	Runtime	Runtime	Compile time

The earlier you catch errors, the more productive you are and the more stable your data processing applications will be.

DataFrame Persistence

A DataFrame can be persisted/cached in memory just like how it is done with RDDs. The same familiar persistence APIs (persist and unpersist) are available in DataFrame class. However, there is one big difference when caching a DataFrame. Since Spark SQL knows the schema of the data in a DataFrame, it can organize the data in a columnar format and apply any applicable compressions to minimize space usage. The net result is it require much less space to store a DataFrame in memory than storing an RDD when both are backed by the same data file. All the different storage options described in Table 3-5 are applicable for persisting a DataFrame. Listing 3-49 demonstrates persisting a DataFrame with a human readable name, which is easy to identify in Spark UI.

Listing 3-49. Persisting a DataFrame with a Human Readable Name

```
val numDF = spark.range(1000).toDF("id")
// register as a view
numDF.createOrReplaceTempView("num_df")
// use Spark catalog to cache the numDF using name "num_df"
spark.catalog.cacheTable("num_df")
// force the persistence to happen by taking the count action
numDF.count
```

Next, point your browser to the Spark UI (http://localhost:4040 when running Spark shell) and click the Storage tab. Figure 3-2 shows an example.

Figure 3-2. *Storage tab*

Summary

In this chapter, you learned the following.

- The Spark SQL module provides a new and powerful abstraction for structured distributed data manipulation. Structured data has a defined schema, which consists of column names and a column data type.

- The main programming abstraction in Spark SQL is the Dataset, and it has two flavors of APIs: a strongly typed API and an untyped API. For the strongly typed APIs, each row is represented by a domain-specified object. For the untyped APIs, the reach row is represented by a Row object. A DataFrame is now just an alias of Dataset[Row]. The strongly-typed APIs give you static-typing and compile-time checking; therefore, they are only available in strongly typed languages, such as Scala or Java.

- Spark SQL supports reading data from a variety of popular data sources in different formats. The `DataFrameReader` class is responsible for creating a DataFrame by reading data from any of these data sources.

- Like RDD, a Dataset has two types of structured operations. They are transformation and actions. The former is lazily evaluated, and the latter is eagerly evaluated.

- Spark SQL makes it very easy to use SQL to perform data processing against large sets. This opens up Spark to data analysts and nonprogrammers.

- Writing out data from either a Dataset or DataFrame is done via a class called `DataFrameWriter`.

SPARK SQL EXERCISES

The following exercises are based on the `movies.tsv` and `movie-ratings.tsv` files in `chapter3/data/movies` directory. The column delimiter in these files is a tab, so make sure to use that splitting each line.

Each line in the `movies.tsv` file represents an actor played in a movie. If a movie has ten actors played in it, there are rows for that movie.

1. Compute the number of movies produced each year. The output should have two columns: year and count. The output should be ordered by the count in descending order.

2. Compute the number of movies each actor was in. The output should have two columns: actor, count. The output should be ordered by the count in descending order.

3. Compute the highest-rated movie per year and include all the actors played in that movie. The output should have only one movie per year, and it should contain four columns: year, movie title, rating, a semicolon-separated list of actor names. This question requires a join between `movies.tsv` and `movie-ratings.tsv` files. There are two approaches to this problem. The first is to figure out the highest-rated movies per year and then join with a list of actors. The second one is to perform the join first and then figure out the highest-rated movies per year and a list of actors. The result of each approach is different from the other one. Why do you think that is?

4. Determine which pair of actors worked together most. Working together is defined as appearing in the same movie. The output should have three columns: actor1, actor2, and count. The output should be sorted by the count in descending order. The solution to this question requires doing self-join.

CHAPTER 4

Spark SQL: Advanced

Chapter 3 introduced the foundational elements in the Spark SQL module, including the core abstraction, structured operations for manipulating structured data, and various supported data sources to read data from and write data to. Building on top of that foundation, this chapter covers some of the advanced capabilities in the Spark SQL module and peeks behind the curtain to understand the optimization and execution efficiency that the Catalyst optimizer and Tungsten engine provide. To help you with performing complex analytics, Spark SQL provides a set of powerful and flexible aggregation capabilities, the ability to join with multiple datasets, a large set of built-in and high-performant functions, an easy way to write your own custom function, and a set of advanced analytic functions. This chapter covers each of these topics in detail.

Aggregations

Performing any interesting and complex analytics on big data usually involves aggregation to summarize the data to extract patterns or insights or generate summary reports. Aggregations usually require grouping either on the entire dataset or based on one or more columns, and then apply aggregation functions such as summation, counting, or average to each group. Spark provides many commonly used aggregation functions and the ability to aggregate the values into a collection, which can then be further analyzed. The grouping of rows can be done at different levels, and Spark supports the following levels.

- Treat a DataFrame as one group.

- Divide a DataFrame into multiple groups using one or more columns and perform one or more aggregations on each group.

- Divide a DataFrame into multiple windows and perform moving average, cumulative sum, or ranking. If a window is based on time, the aggregations can be done per tumbling or sliding windows.

H. Luu, *Beginning Apache Spark 3*, https://doi.org/10.1007/978-1-4842-7383-8_4

Aggregation Functions

In Spark, all aggregations are done via functions. The aggregation functions are designed to perform aggregation on a set of rows, whether rows consist of all the rows or a subgroup of rows in a DataFrame. The documentation of the complete list of aggregation functions for the Scala language is available at http://spark.apache.org/docs/latest/api/scala/index.html#org.apache.spark.sql.functions$. For the Spark Python APIs, sometimes there are some gaps in the availability of some functions.

Common Aggregation Functions

This section describes a set of commonly used aggregation functions and provides examples of working with them. Table 4-1 describes the aggregation function. For a complete list, please see http://spark.apache.org/docs/latest/api/scala/index.html#org.apache.spark.sql.functions$.

Table 4-1. *Commonly Used Aggregation Functions*

Operation	Description
count(col)	Return the number of items per group.
countDistinct(col)	Return the unique number of items per group.
approx_count_distinct(col)	Return the approximate number of unique items per group.
min(col)	Return the minimum value of the given column per group.
max(col)	Return the maximum value of the given column per group.
sum(col)	Return the sum of the values in the given column per group.
sumDistinct(col)	Return the sum of the distinct values of the given column per group.
avg(col)	Return the average of the values of the given column per group.
skewness(col)	Return the skewness of the distribution of the values of the given column per group.
kurtosis(col)	Return the kurtosis of the distribution of the values of the given column per group.
variance(col)	Return the unbiased variance of the values of the given column per group.

(continued)

Table 4-1. (*continued*)

Operation	Description
stddev(col)	Return the standard deviation of the values of the given column per group.
collect_list(col)	Return a collection of values of the given column. The returned collection may contain duplicate values.
collect_set(col)	Return a collection of unique values of the given column.

To demonstrate the usage of these functions, let's use the flight summary dataset, which is derived from the data files available on the Kaggle site at www.kaggle.com/ usdot/flight-delays/data. This dataset contains the 2015 US domestic flight delays and cancellations. Listing 4-1 is the code for creating a DataFrame from this dataset.

Listing 4-1. Create a DataFrame from Reading Flight Summary Dataset

```
val flight_summary = spark.read.format("csv")
                                    .option("header", "true")
                                    .option("inferSchema","true")
                                    .load("<path>/chapter5/data/
flights/flight-summary.csv")
// use count action to find out number of rows in this dataset
flight_summary.count()
Long = 4693
Remember the count() function of the DataFrame is an action so it
immediately returns a value to us. All the functions listed in Table 5-1
are lazily evaluated functions.
Below is the schema of the flight_summary dataset.
 |-- origin_code: string (nullable = true)
 |-- origin_airport: string (nullable = true)
 |-- origin_city: string (nullable = true)
 |-- origin_state: string (nullable = true)
 |-- dest_code: string (nullable = true)
 |-- dest_airport: string (nullable = true)
 |-- dest_city: string (nullable = true)
 |-- dest_state: string (nullable = true)
 |-- count: integer (nullable = true)
```

Each row represents the flights from the origin_airport to dest_airport. The count column has the number of flights.

All the aggregation examples below are performing aggregation at the entire DataFrame level. Examples of performing aggregations at the subgroups level are given later in the chapter.

count(col)

Counting is a commonly used aggregation to find out the number of items in a group. Listing 4-2 computes the count for both the origin_airport and dest_airport columns, and as expected, the count is the same. To improve the readability of the result column, you can use the as function to give a friendlier column name. Note that you need to call the show action to see the result.

Listing 4-2. Computing the Count for Two Columns in the flight_summary DataFrame

```
flight_summary.select(count("origin_airport"), count("dest_airport").
as("dest_count")).show
+--------------------------+---------------+
|     count(origin_airport)|     dest_count|
+--------------------------+---------------+
|                      4693|           4693|
+--------------------------+---------------+
```

When counting the number of items in a column, the count(col) function doesn't include the null value in the count. To include the null value, the column name should be replaced with *. Listing 4-3 demonstrates this behavior by creating a small DataFrame with a null value in some columns.

Listing 4-3. Counting Items with Null Value

```
import org.apache.spark.sql.Row
case class Movie(actor_name:String, movie_title:String, produced_year:Long)
val badMoviesDF = Seq( Movie(null, null, 2018L),
                       Movie("John Doe", "Awesome Movie", 2018L),
                       Movie(null, "Awesome Movie", 2018L),
                       Movie("Mary Jane", "Awesome Movie", 2018L)).toDF
```

```
badMoviesDF.show
+--------------+--------------------+--------------+
|    actor_name|         movie_title| produced_year|
+--------------+--------------------+--------------+
|          null|                null|          2018|
|      John Doe|       Awesome Movie|          2018|
|          null|       Awesome Movie|          2018|
|     Mary Jane|       Awesome Movie|          2018|
+--------------+--------------------+--------------+

// now performing the count aggregation on different columns
badMoviesDF.select(count("actor_name"), count("movie_title"),
count("produced_year"), count("*")).show
+------------------+-------------------+--------------------+--------+
| count(actor_name)| count(movie_title)| count(produced_year)| count(1)|
+------------------+-------------------+--------------------+--------+
|                 2|                  3|                   4|       4|
+------------------+-------------------+--------------------+--------+
```

The output table confirms that the count(col) function doesn't include null the in the final count.

countDistinct(col)

This function does what it sounds like. It only counts the unique items per group. Listing 4-4 shows the differences in the count result between the countDistinct function and the count function. As it turns out, there are 322 unique airports in the flight_summary dataset.

Listing 4-4. Counting Unique Items in a Group

```
flight_summary.select(countDistinct("origin_airport"), countDistinct("dest_
airport"), count("*")).show
+---------------------------+-------------------------+--------+
| count(DISTINCT origin_airport)| count(DISTINCT dest_airport)| count(1)|
+---------------------------+-------------------------+--------+
|                        322|                      322|    4693|
+---------------------------+-------------------------+--------+
```

approx_count_distinct (col, max_estimated_error=0.05)

Counting the exact number of unique items in each group in a very large dataset is an expensive and time-consuming. In some use cases, it is sufficient to have an approximate unique count. One of those use cases is in the online advertising business, where there are hundreds of millions of ad impressions per hour. There is a need to generate a report on the number of unique visitors per certain type of member segment. Approximating a count of distinct items is a well-known problem in computer science. It is also known as the *cardinality estimation problem.*

Luckily, there is already a well-known algorithm called HyperLogLog (https:// en.wikipedia.org/wiki/HyperLogLog) that you can use to solve this problem, and Spark has implemented a version of this algorithm in the approx_count_distinct function. Since the unique count is an approximation, there is a certain amount of error. This function allows you to specify the value for an acceptable estimation error for this use case. Listing 4-5 demonstrates the usage and behavior of the approx._count_ distinct function. As you dial down the estimation error, it takes longer and longer for this function to complete and return the result.

Listing 4-5. Counting Unique Items in a Group

```
// let's do the counting on the "count" column of flight_summary DataFrame.
// the default estimation error is 0.05 (5%)
flight_summary.select(count("count"),countDistinct("count"), approx_count_
distinct("count", 0.05)).show
+--------------+----------------------+----------------------------+
| count(count) | count(DISTINCT count)| approx_count_distinct(count)|
+--------------+----------------------+----------------------------+
|          4693|                  2033|                        2252|
+--------------+----------------------+----------------------------+

// to get a sense how much approx_count_distinct function is faster than
countDistinct function,
// trying calling them separately
flight_summary.select(countDistinct("count")).show

// specify 1% estimation error
flight_summary.select(approx_count_distinct("count", 0.01)).show
```

// one my Mac laptop, the approx_count_distinct function took about
0.1 second and countDistinct function took 0.6 second. The larger the
approximation estimation error, the less time approx_count_distinct
function takes to complete.

min(col), max(col)

The minimum and maximum values of the items in a group are the two ends of a
spectrum. These two functions are easy to understand and work with. Listing 4-6 extracts
these two values from the count column.

Listing 4-6. Get the Minimum and Maximum Values of the Count Column

```
flight_summary.select(min("count"), max("count")).show
```

```
+-------------+----------------+
|   min(count)|      max(count)|
+-------------+----------------+
|            1|           13744|
+-------------+----------------+
```

// looks like there is one very busy airport with 13744 incoming flights
from another airport. It will be interesting to find which airport

sum(col)

This function computes the sum of the values in a numeric column. Listing 4-7 performs
the sum of all the flights in the flight_summary dataset.

Listing 4-7. Using sum Function to Sum up the Count Values

```
flight_summary.select(sum("count")).show
```

```
+---------------+
|    sum(count)|
+---------------+
|        5332914|
+---------------+
```

sumDistinct(col)

This function does what it sounds like. It sums up only the distinct values of a numeric column. The sum of the distinct counts in the `flight_summary` DataFrame should be less than the total sum displayed in Listing 4-7. Listing 4-8 computes the sum of the distinct values.

Listing 4-8. Using sumDistinct Function to Sum up the Distinct Count Values

```
flight_summary.select(sumDistinct("count")).show
+---------------------------+
|         sum(DISTINCT count)|
+---------------------------+
|                     3612257|
+---------------------------+
```

avg(col)

This function calculates the average value of a numeric column. This convenient function simply takes the total and divides it by the number of items. Let's see whether Listing 4-9 can validate the hypothesis.

Listing 4-9. Computing the Average Value of the Count Column Using Two Different Ways

```
flight_summary.select(avg("count"), (sum("count") / count("count"))).show

+-----------------------+-----------------------------------+
|            avg(count)|         (sum(count) / count(count))|
+-----------------------+-----------------------------------+
|      1136.3549968037503|                 1136.3549968037503|
+-----------------------+-----------------------------------+
```

skewness(col), kurtosis(col)

In statistics, the distribution of the values in a dataset tells numerous stories behind the dataset. Skewness measures the symmetry of the value distribution in a dataset, and its value can be positive, zero, negative, or undefined. In a normal distribution or bell-shaped distribution, the skew value is 0. A positive skew indicates the tail on the right side is longer or fatter than the left side. A negative skew indicates the opposite, where the tail of the left side is longer or fatter than the right side. The tail of both sides is even when the skew is 0. Figure 4-1 shows an example of negative and positive skew.

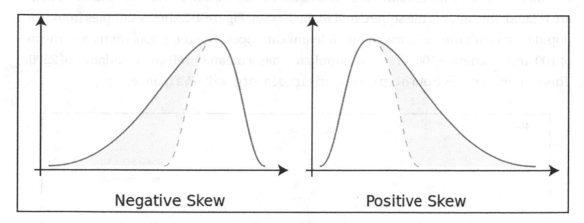

Figure 4-1. Negative and positive skew examples from https://en.wikipedia.org/wiki/Skewness

Kurtosis is a measure of the shape of the distribution curve, whether the curve is normal, flat, or pointy. Positive kurtosis indicates the curve is slender and pointy, and negative kurtosis indicates fat and flat. Listing 4-10 calculates the skewness and kurtosis for the count distribution in the flight_summary dataset.

Listing 4-10. Compute the Skewness and Kurtosis of Column Count

```
flight_summary.select(skewness("count"), kurtosis("count")).show
+--------------------------+---------------------------+
|         skewness(count)|           kurtosis(count)|
+--------------------------+---------------------------+
|        2.682183800064101|         10.51726963017102|
+--------------------------+---------------------------+
```

The result suggests the distribution of the counts is not symmetric, and the right tail is longer or fatter than the left tail. The kurtosis value suggests that the distribution curve is pointy.

variance(col), stddev(col)

In statistics, variance, and standard deviation measure the dispersion or the spread of the data. In other words, they tell the average distance of the values from the mean. When the variance value is low, the values are close to the mean. Variance and standard deviation are related; the latter is the square root of the former. Figure 4-2 shows samples from two populations with the same mean but different variances. The red population has a mean of 100 and a variance 100. The blue population has a mean of 100 and a variance of 2500. This example comes from `https://en.wikipedia.org/wiki/Variance`.

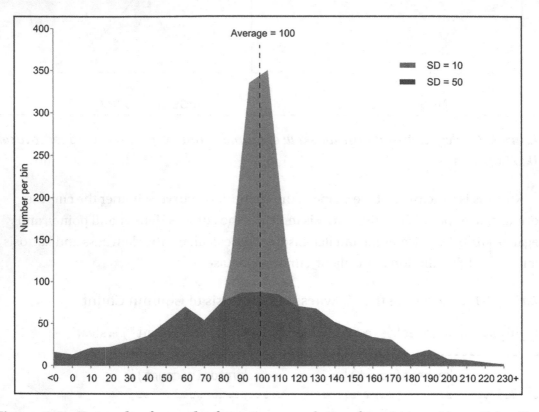

Figure 4-2. *Example of samples from two population from* `https://en.wikipedia.` `org/wiki/Variance`

The `variance` and `stddev` calculate the variance and standard deviation, respectively. Spark provides two different implementations of these functions; one uses sampling to speed up the calculation, and the other uses the entire population. Listing 4-11 shows the variance and standard deviation of the count column in the `flight_summary` DataFrame.

Listing 4-11. Compute the Variance and Standard Deviation Using variance and sttdev Functions

```
// use the two variations of variance and standard deviation
flight_summary.select(variance("count"), var_pop("count"), stddev("count"),
stddev_pop("count")).show
```

```
+----------------+-----------------+------------------+-----------------+
|  var_samp(count)|      var_pop(count)| stddev_samp(count)| stddev_pop(count)|
+----------------+-----------------+------------------+-----------------+
|1879037.7571558713| 1878637.3655604832|  1370.779981308405| 1370.633928355957|
+----------------+-----------------+------------------+-----------------+
```

It looks like the count values are pretty spread out in `flight_summary` DataFrame.

Aggregation with Grouping

This section covers aggregation with the grouping of one or more columns. The aggregations are usually performed on datasets that contain one or more categorical columns, which have low cardinality. Examples of categorical values are gender, age, city name, or country name. The aggregation is done through functions similar to the ones mentioned earlier. However, instead of performing aggregation on the global group in the DataFrame, they perform the aggregation on each subgroup.

Performing aggregation with grouping is a two-step process. The first step is to perform the grouping by using the `groupBy(col1,col2,...)` transformation, and that's where you specify which columns to group the rows by. Unlike other transformations that return a DataFrame, the `groupBy` transformation returns an instance of `RelationalGroupedDataset` class, to which you can apply one or more aggregation functions. Listing 4-12 demonstrates a simple grouping of using one column and one aggregation. Notice the `groupBy` columns automatically be included in the output.

Listing 4-12. Grouping by origin_airport and Perform Count Aggregation

```
flight_summary.groupBy("origin_airport").count().show(5, false)
+----------------------------------------------------------+-------+
|                  origin_airport                          | count|
+----------------------------------------------------------+-------+
|Melbourne International Airport                            |      1|
|San Diego International Airport (Lindbergh Field)          |     46|
|Eppley Airfield                                           |     21|
|Kahului Airport                                           |     18|
|Austin-Bergstrom International Airport                     |     41|
+----------------------------------------------------------+-------+
```

Listing 4-12 shows the flights out of Melbourne International Airport (Florida) go to only one other airport. However, the flights out of the Kahului Airport land at one of 18 other airports.

To make things a bit more interesting, let's try grouping by two columns to calculate the same metric at the city level. Listing 4-13 shows how to do that.

Listing 4-13. Grouping by origin_state and origin_city and Perform Count Aggregation

```
flight_summary.groupBy('origin_state, 'origin_city).
count().                   .where('origin_state === "CA").orderBy('count.
desc).show(5)
```

```
+---------------+------------------+---------+
|  origin_state|        origin_city|   count|
+---------------+------------------+---------+
|            CA|     San Francisco|      80|
|            CA|       Los Angeles|      80|
|            CA|         San Diego|      47|
|            CA|           Oakland|      35|
|            CA|        Sacramento|      27|
+---------------+------------------+---------+
```

In addition to grouping by two columns, the statement filters the rows to only the ones with a "CA" state. The orderBy transformation makes it easy to identify which city has the greatest number of destination airports. It makes sense that both San Francisco and Los Angeles in California have the largest number of destination airports that one can fly to.

The RelationalGroupedDataset class provides a standard set of aggregation functions that you can use to apply to each subgroup. They are avg(cols), count(), mean(cols), min(cols), max(cols), sum(cols). Except for the count() function, all the remaining ones operate on numeric columns.

Multiple Aggregations per Group

Sometimes there is a need to perform multiple aggregations per group at the same time. For example, in addition to the count, you want to know the minimum and maximum values. The RelationalGroupedDataset class provides a very powerful function called agg that takes one or more column expressions, which means you can use any of the aggregation functions, including the ones listed in Table 4-1. One cool thing is these aggregation functions return an instance of the Column class, so you can then apply any of the column expressions using the provided functions. A common need is to rename the column after the aggregation is done to make it shorter, more readable, and easier to refer to. Listing 4-14 demonstrates how to do all of this.

Listing 4-14. Multiple Aggregations After a Group by of origin_airport

```
import org.apache.spark.sql.functions._
flight_summary.groupBy("origin_airport")
                    .agg(
                            count("count").as("count"),
                            min("count"), max("count"),
                            sum("count")
                    ).show(5)
+--------------------+------+----------+----------+------------+
|     origin_airport| count|min(count)|max(count)| sum(count)|
+--------------------+------+----------+----------+------------+
|Melbourne Interna...|     1|      1332|      1332|        1332|
|San Diego Interna...|    46|         4|      6942|       70207|
|     Eppley Airfield|    21|         1|      2083|       16753|
|     Kahului Airport|    18|        67|      8313|       20627|
|Austin-Bergstrom ...|    41|         8|      4674|       42067|
+--------------------+------+----------+----------+------------+
```

By default, the aggregation column name is the aggregation expression, making the column name a bit long and difficult to refer to. Therefore, a common pattern is to use the `Column.as` function to rename the column to something more suitable.

The versatile `agg` function provides an additional way to express the column expressions via a string-based key-value map. The key is the column name, and the value is the aggregation method, which can be `avg, max, min, sum,` or `count.` Listing 4-15 provides an example of this approach.

Listing 4-15. Specifying Multiple Aggregations Using a Key-Value Map

```
flight_summary.groupBy("origin_airport")
                  .agg(
                            "count" -> "count",
                            "count" -> "min",
                            "count" -> "max",
                            "count" -> "sum")
              .show(5)
```

The result is the same as the one from Listing 4-14. Notice there isn't an easy to rename the aggregation result column name. One advantage this approach has over the first one is the map can programmatically be generated. When writing production ETL jobs or performing exploratory analysis, the first approach is used more often than the second one.

Collection Group Values

The `collect_list(col)` and `collect_set(col)` functions are useful to collect all the values of a particular group after the grouping is applied. Once the values of each group are placed in a collection, there is freedom to operate them any way you choose. There is one small difference in the returned collection of these functions, which is the uniqueness. The `collection_list` function returns a collection containing duplicate values, and the `collection_set` function returns a collection containing unique values. Listing 4-16 shows using the `collection_list` function to collect the destination cities with more than 5500 flights coming into them from each origin state.

Listing 4-16. Using collection_list to Collect High Traffic Destination Cities Per Origin State

```
val highCountDestCities = flight_summary.where('count > 5500)
                              .groupBy("origin_state")
                              .agg(collect_list("dest_city")
                              .as("dest_cities"))
highCountDestCities.withColumn("dest_city_count",
                              size('dest_cities))
                .show(5, false)
+------------+-------------------------------------+---------------+
|origin_state|           dest_cities               | dest_city_count|
+------------+-------------------------------------+---------------+
|          AZ|      [Seattle, Denver, Los Angeles]|              3|
|          LA|      [Atlanta]                     |              1|
|          MN|      [Denver, Chicago]             |              2|
|          VA|      [Chicago, Boston, Atlanta]    |              3|
|          NV|[Denver, Los Angeles, San Francisco]|              3|
+------------+-------------------------------------+---------------+
```

Aggregation with Pivoting

Pivoting is a way to summarize the data by specifying one of the categorical columns and then performing aggregations on other columns so that the categorical values are transposed from rows into individual columns. Another way of thinking about pivoting is that it is a way to translate rows into columns while applying one or more aggregations. This technique is commonly used in data analysis or reporting. The pivoting process starts with grouping one or more columns, pivots on a column, and finally ends with applying one or more aggregations on one or more columns.

Listing 4-17 shows a pivoting example on a small dataset of students where each row contains the student's name, gender, weight, and graduation year. Pivoting makes it easy to compute the average weight of each gender for each graduation year.

Listing 4-17. Pivoting on a Small Dataset

```
import org.apache.spark.sql.Row

case class Student(name:String, gender:String, weight:Int, graduation_
year:Int)

val studentsDF = Seq(Student("John", "M", 180, 2015),
                     Student("Mary", "F", 110, 2015),
                     Student("Derek", "M", 200, 2015),
                     Student("Julie", "F", 109, 2015),
                     Student("Allison", "F", 105, 2015),
                     Student("kirby", "F", 115, 2016),
                     Student("Jeff", "M", 195, 2016)).toDF

// calculating the average weight for gender per graduation year
studentsDF.groupBy("graduation_year").pivot("gender")
                                     .avg("weight").show()

+----------------+------+---------+
| graduation_year|     F|        M|
+----------------+------+---------+
|            2015| 108.0|    190.0|
|            2016| 115.0|    195.0|
+----------------+------+---------+
```

This example has only one aggregation, and the gender categorical column has only two possible unique values; therefore, the result table has only two columns. If the gender column has three possible unique values, there are three columns in the result table. You can leverage the `agg` function to perform multiple aggregations, creating more columns in the result table. Listing 4-18 is an example of performing multiple aggregations on the DataFrame from Listing 4-17.

Listing 4-18. Multiple Aggregations After Pivoting

```
studentsDF.groupBy("graduation_year").pivot("gender")
                  .agg(
                          min("weight").as("min"),
                          max("weight").as("max"),
```

```
                    avg("weight").as("avg")
            ).show()
```

```
+---------------+------+------+------+------+------+------+
|graduation_year| F_min| F_max| F_avg| M_min| M_max| M_avg|
+---------------+------+------+------+------+------+------+
|           2015|   105|   110| 108.0|   180|   200| 190.0|
|           2016|   115|   115| 115.0|   195|   195| 195.0|
+---------------+------+------+------+------+------+------+
```

The number of columns added after the group columns in the result table is the product of the number of unique values of the pivot column and the number of aggregations.

If the pivoting column has a lot of distinct values, you can selectively choose which values to generate the aggregations for. Listing 4-19 shows how to specify values to the pivoting function.

Listing 4-19. Selecting Values of Pivoting Column to Generate the Aggregations For

```
studentsDF.groupBy("graduation_year").pivot("gender", Seq("M"))
            .agg(
                    min("weight").as("min"),
                    max("weight").as("max"),
                    avg("weight").as("avg")
            ).show()
```

```
+--------------------+---------+----------+---------+
|     graduation_year|    M_min|     M_max|    M_avg|
+--------------------+---------+----------+---------+
|                2015|      180|       200|    190.0|
|                2016|      195|       195|    195.0|
+--------------------+---------+----------+---------+
```

Specifying a list of distinct values for the pivot column speeds up the pivoting process. Otherwise, Spark spends some effort in figuring out a list of distinct values on its own.

Joins

To perform any kind of complex and interesting data analysis or manipulations, you often need to bring together the data from multiple datasets through the process of joining. This is a well-known technique in SQL parlance. Performing a join combines the columns of two datasets (could be different or same), and the combined dataset contains columns from both sides. This enables you to further analyze the combined dataset so that it is not possible with each set. Let's take an example of the two datasets from an online e-commerce company. One represents the transactional data that contains information about which customers purchased what products (a.k.a. fact table). The other one represents the information on each customer (a.k.a. dimension table). By joining these two datasets, you can extract insights about which products are more popular with certain segments of customers in terms of age or location.

This section covers how to perform joining in Spark SQL using the `join` transformation and the various types of join it supports. The last portion of this section describes how Spark SQL internally performs the joining.

Note   In the world of performing data analysis using SQL, a join is a technique used quite often. If you are new to SQL, it is highly recommended that you learn the fundamental concepts and the different kinds of join at https://en.wikipedia. org/wiki/Join_(SQL). A few tutorials about joins are provided at `www.w3schools. com/sql/sql_join.asp`.

Join Expression and Join Types

Performing a join of two datasets requires you to specify two pieces of information. The first one is a join expression that specifies which columns from each side should determine which rows from both datasets are included in the joined dataset. The second one is the join type, which determines what should be included in the joined dataset. Table 4-2 provides a list of supported join types in Spark SQL.

Table 4-2. *Join Types*

Type	Description
Inner join (a.k.a. equi-join)	Return rows from both datasets when the join expression evaluates to true.
Left outer join	Return rows from the left dataset even when the join expression evaluates as false.
Right outer join	Return rows from the right dataset even when the join expression evaluates as false.
Outer join	Return rows from both datasets even when the join expression evaluates as false.
Left anti-join	Return rows only from the left dataset when the join expression evaluates as false.
Left semi-join	Return rows only from the left dataset when the join expression evaluates to true.
Cross (a.k.a. Cartesian)	Return rows by combining each row from the left dataset with each row in the right dataset. The number of rows is a product of the size of each dataset.

To help visualize some of the join types, Figure 4-3 shows a set of Venn diagrams for the common join types from https://en.wikipedia.org/wiki/Join_ (SQL)#Outer_join.

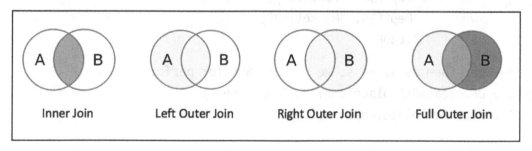

Figure 4-3. *Venn diagrams for common join types*

Working with Joins

I used two small DataFrames to demonstrate how to perform joining in Sparking SQL. The first one represents a list of employees, and each row contains the employee's name and the department they belong to. The second one contains a list of departments, and each row contains a department ID and department name. Listing 4-20 contains a snippet of code to create these two DataFrames.

Listing 4-20. Creating Two Small DataFrames to Use in the Following Join Type Examples

```
case class Employee(first_name:String, dept_no:Long)
val employeeDF = Seq( Employee("John", 31),
                      Employee("Jeff", 33),
                      Employee("Mary", 33),
                      Employee("Mandy", 34),
                      Employee("Julie", 34),
                      Employee("Kurt", null.asInstanceOf[Int])
                    ).toDF

case class Dept(id:Long, name:String)
val deptDF = Seq( Dept(31, "Sales"),
                  Dept(33, "Engineering"),
                  Dept(34, "Finance"),
                  Dept(35, "Marketing")
                ).toDF

// register them as views so we can use SQL for perform joins
employeeDF.createOrReplaceTempView("employees")
deptDF.createOrReplaceTempView("departments")
```

Inner Joins

This is the most used join type with the join expression containing the equality comparison of the columns from both datasets. The joined dataset contains the rows only when the join expression is evaluated to be true; in other words, the join column values are the same in both datasets. Rows that don't have matching column values are excluded from the joined dataset. If the join expression uses the equality comparison, then the number of rows in the joined table only be as large as the size of the smaller dataset. The inner join is the default join type in Spark SQL, so it is optional to specify it in the join transformation. Listing 4-21 provides examples of doing an inner join.

Listing 4-21. Performing Inner Join by the Department ID

```
// define the join expression of equality comparison
val deptJoinExpression = employeeDF.col("dept_no") === deptDF.col("id")

// perform the join
employeeDF.join(deptDF, joinExpression, "inner").show

// no need to specify the join type since "inner" is the default
employeeDF.join(deptDF, joinExpression).show
+-------------+---------+---+----------------+
|   first_name|  dept_no| id|            name|
+-------------+---------+---+----------------+
|         John|       31| 31|           Sales|
|         Jeff|       33| 33|     Engineering|
|         Mary|       33| 33|     Engineering|
|        Mandy|       34| 34|         Finance|
|        Julie|       34| 34|         Finance|
+-------------+---------+---+----------------+

// using SQL
spark.sql("select * from employees JOIN departments on dept_no == id").show
```

As expected, the joined dataset contains only the rows with matching department IDs from both employee and department datasets and the columns from both datasets. The output tells you exactly which department each employee belongs to.

The join expression can be specified in the join transformation or using the where transformation. If the column names are unique, it is possible to refer to the columns in the join expression using a short-handed version. If not, you must specify which DataFrame a particular column comes from using the col function. Listing 4-22 shows different ways of expressing a join expression.

Listing 4-22. Different Ways of Expressing a Join Expression

```
// a shorter version of the join expression
employeeDF.join(deptDF, 'dept_no === 'id).show

// specify the join expression inside the join transformation
employeeDF.join(deptDF, employeeDF.col("dept_no") === deptDF.col("id")).show

// specify the join expression using the where transformation
employeeDF.join(deptDF).where('dept_no === 'id).show
```

Join expression is simply a Boolean predicate, and therefore it can be as simple as comparing two columns or as complex as chaining multiple logical comparisons of pairs of columns.

Left Outer Joins

The joined dataset of this join type includes all the rows from an inner join plus all the rows from the left dataset that the join expression is evaluated as false. For those nonmatching rows, it fills in a NULL value for the columns of the right dataset. Listing 4-23 is an example of doing a left outer join.

Listing 4-23. Performing a Left Outer Join

```
// the join type can be either "left_outer" or "leftouter"
employeeDF.join(deptDF, 'dept_no === 'id, "left_outer").show

// using SQL
spark.sql("select * from employees LEFT OUTER JOIN departments on dept_no
== id").show
```

```
+--------------+----------+----+---------------+
|    first_name|   dept_no|  id|           name|
+--------------+----------+----+---------------+
|          John|        31|  31|          Sales|
|          Jeff|        33|  33|    Engineering|
|          Mary|        33|  33|    Engineering|
|         Mandy|        34|  34|        Finance|
|         Julie|        34|  34|        Finance|
|          Kurt|         0|null|           null|
+--------------+----------+----+---------------+
```

As expected, the marketing department doesn't have any matching rows from the employee dataset. The joined dataset tells you the department that an employee is assigned to and which departments have no employees.

Right Outer Joins

The behavior of this join type resembles the behavior of the left outer join type, except the same treatment is applied to the right dataset. In other words, the joined dataset includes all the rows from an inner join plus all the rows from the right dataset that the join expression evaluates as false. Listing 4-24 is an example of doing a right outer join.

Listing 4-24. Performing a Right Outer Join

```
employeeDF.join(deptDF, 'dept_no === 'id, "right_outer").show

// using SQL
spark.sql("select * from employees RIGHT OUTER JOIN departments on dept_no
== id").show
+-------------+----------+----+----------------+
|   first_name|   dept_no|  id|            name|
+-------------+----------+----+----------------+
|         John|        31|  31|           Sales|
|         Mary|        33|  33|     Engineering|
|         Jeff|        33|  33|     Engineering|
|        Julie|        34|  34|         Finance|
|        Mandy|        34|  34|         Finance|
|         null|      null|  35|       Marketing|
+-------------+----------+----+----------------+
```

As expected, the marketing department doesn't have any match rows from the employee dataset. The joined dataset tells you the department that an employee is assigned to and which departments have no employees.

Outer Joins (a.k.a. Full Outer Joins)

The behavior of this join type is effectively the same as combining the result of both the left outer join and the right outer join. Listing 4-25 is an example of doing an outer join.

Listing 4-25. Performing an Outer Join

```
employeeDF.join(deptDF, 'dept_no === 'id, "outer").show

// using SQL
spark.sql("select * from employees FULL OUTER JOIN departments on dept_no
== id").show
```

```
+------------+----------+----+---------------+
| first_name| dept_no| id|           name|
+------------+----------+----+---------------+
|        Kurt|        0|null|           null|
|       Mandy|       34|  34|        Finance|
|       Julie|       34|  34|        Finance|
|        John|       31|  31|          Sales|
|        Jeff|       33|  33|    Engineering|
|        Mary|       33|  33|    Engineering|
|        null|     null|  35|      Marketing|
+------------+----------+----+---------------+
```

The result from the outer join allows you to see which department an employee is assigned to and which departments have employees and which employees are not assigned to a department and which departments don't have any employees.

Left Anti-Joins

This join type lets you find out which rows from the left dataset don't have any matching rows on the right dataset, and the joined dataset contains only the columns from the left dataset. Listing 4-26 is an example of doing a left anti-join.

Listing 4-26. Performing a Left Anti-Join

```
employeeDF.join(deptDF, 'dept_no === 'id, "left_anti").show

// using SQL
spark.sql("select * from employees LEFT ANTI JOIN departments on dept_no ==
id").show
```

```
+-------------+----------+
|   first_name|   dept_no|
+-------------+----------+
|         Kurt|         0|
+-------------+----------+
```

The result from the left anti-join can easily tell you which employees are not assigned to a department. Notice the right anti-join type doesn't exist; however, you can easily switch the datasets around to achieve the same goal.

Left Semi-Joins

The behavior of this join type is similar to the inner join type, except the joined dataset doesn't include the columns from the right dataset. Another way of thinking about this join type is its behavior is the opposite of the left anti-join, where the joined dataset contains only the matching rows. Listing 4-27 is an example of doing a left semi-join.

Listing 4-27. Performing a Left Semi-Join

```
employeeDF.join(deptDF, 'dept_no === 'id, "left_semi").show

// using SQL
spark.sql("select * from employees LEFT SEMI JOIN departments on dept_no ==
id").show
+-------------+----------+
|   first_name|   dept_no|
+-------------+----------+
|         John|        31|
|         Jeff|        33|
|         Mary|        33|
|        Mandy|        34|
|        Julie|        34|
+-------------+----------+
```

Cross (a.k.a. Cartesian)

In terms of usage, this join type is the simplest to use because the join expression is not needed. Its behavior can be a bit dangerous because it joins every single row in the left dataset with every row in the right dataset. The size of the joined dataset is the product of the size of the two datasets. For example, if each dataset size is 1024, then the size of the joined dataset is over 1 million rows. For this reason, the way to use this join type is by explicitly using a dedicated transformation in the DataFrame class, rather than specifying this join type as a string. Listing 4-28 is an example of doing a cross join.

Listing 4-28. Performing a Cross Join

```
// using crossJoin transformation and display the count
employeeDF.crossJoin(deptDF).count
Long = 24

// using SQL and passing 30 value to show action to see all rows
spark.sql("select * from employees CROSS JOIN departments").show(30)
+------------+---------+---+----------------+
|  first_name| dept_no| id|            name|
+------------+---------+---+----------------+
|        John|      31| 31|           Sales|
|        John|      31| 33|     Engineering|
|        John|      31| 34|         Finance|
|        John|      31| 35|       Marketing|
|        Jeff|      33| 31|           Sales|
|        Jeff|      33| 33|     Engineering|
|        Jeff|      33| 34|         Finance|
|        Jeff|      33| 35|       Marketing|
|        Mary|      33| 31|           Sales|
|        Mary|      33| 33|     Engineering|
|        Mary|      33| 34|         Finance|
|        Mary|      33| 35|       Marketing|
|       Mandy|      34| 31|           Sales|
|       Mandy|      34| 33|     Engineering|
|       Mandy|      34| 34|         Finance|
|       Mandy|      34| 35|       Marketing|
```

```
|          Julie|        34| 31|           Sales|
|          Julie|        34| 33|     Engineering|
|          Julie|        34| 34|         Finance|
|          Julie|        34| 35|       Marketing|
|           Kurt|         0| 31|           Sales|
|           Kurt|         0| 33|     Engineering|
|           Kurt|         0| 34|         Finance|
|           Kurt|         0| 35|       Marketing|
+--------------+----------+---+----------------+
```

Dealing with Duplicate Column Names

From time to time, two DataFrames might have one or more columns with the same name. Before joining them, it is best to rename those columns in one of the two DataFrames to avoid access ambiguity issues; otherwise, the joined DataFrame would have multiple columns with the same name. Listing 4-29 simulates this situation.

Listing 4-29. Simulate a Joined DataFrame with Multiple Names That Are the Same

```
// add a new column to deptDF with name dept_no
val deptDF2 = deptDF.withColumn("dept_no", 'id)

deptDF2.printSchema
 |-- id: long (nullable = false)
 |-- name: string (nullable = true)
 |-- dept_no: long (nullable = false)

// now employeeDF with deptDF2 using dept_no column
val dupNameDF = employeeDF.join(deptDF2, employeeDF.col("dept_no") ===
deptDF2.col("dept_no"))

dupNameDF.printSchema
 |-- first_name: string (nullable = true)
 |-- dept_no: long (nullable = false)
 |-- id: long (nullable = false)
 |-- name: string (nullable = true)
 |-- dept_no: long (nullable = false)
```

Notice the dupNameDF DataFrame now has two columns with the same name, dept_no. Spark throws an error when you project the dupNameDF DataFrame using the dept_no in Listing 4-30.

Listing 4-30. Projecting Column dept_no in the dupNameDF DataFrame

```
dupNameDF.select("dept_no")

org.apache.spark.sql.AnalysisException: Reference 'dept_no' is ambiguous,
could be: dept_no#30L, dept_no#1050L.;
```

As it turns out, there are several ways to deal with this issue.

Use Original DataFrame

The joined DataFrame remembers which columns come from which original DataFrame during the joining process. To disambiguate which DataFrame a column comes from, you can just tell Spark to prefix it with its original DataFrame name. Listing 4-31 shows how to do this.

Listing 4-31. Using the Original DataFrame deptDF2 to Refer to dept_no Column in the Joined DataFrame

```
dupNameDF.select(deptDF2.col("dept_no"))
```

Renaming Column Before Joining

Another approach to avoid a column name ambiguity issue is to rename a column in one of the DataFrames using the withColumnRenamed transformation. Since this is simple, I leave it as an exercise for you.

Using Joined Column Name

When the joined column name is the same in both DataFrames, you can leverage a version of the join transformation that automatically removes the duplicate column name in the joined DataFrame. However, if it was a self-join, meaning joining a DataFrame to itself, then there is no way to refer to other duplicate column names. In that case, you need to use the column renaming technique. Listing 4-32 shows an example of performing a join using a joined column name.

Listing 4-32. Performing a Join Using Joined Column Name

```
val noDupNameDF = employeeDF.join(deptDF2, "dept_no")

noDupNameDF.printSchema
 |-- dept_no: long (nullable = false)
 |-- first_name: string (nullable = true)
 |-- id: long (nullable = false)
 |-- name: string (nullable = true)
```

Notice there is only one dept_no column in the noDupNameDF DataFrame.

Overview of Join Implementation

Joining is one of the most complex and expensive operations in Spark. At a high level, there are a few strategies Spark uses to perform the joining of two datasets. They are shuffle hash join and broadcast join. The main criteria for selecting a particular strategy are based on the size of the two datasets. When the size of both datasets is large, then the shuffle hash join strategy is used. When the size of one of the datasets is small enough to fit into the memory of the executor, then the broadcast join strategy is used. The following sections go into detail on how each joining strategy works.

Shuffle Hash Join

Conceptually, joining is about combining the rows of two datasets that meet the condition in the join expression. To do that, rows with the same column values need to be transferred across the network, co-located on the same partition.

The shuffle hash join implementation consists of two steps. The first step computes the hash value of the column(s) in the join expression of each row in each dataset and then shuffles those rows with the same hash value to the same partition. To determine which partition a particular row is moved to, Spark performs a simple arithmetic operation, which computes the modulo of the hash value by the number of partitions. The second step combines the columns of those rows that have the same column hash value. At the high level, these two steps are like the steps in the MapReduce programming model.

Figure 4-4 shows the shuffling going on in the shuffle hash join. It is an expensive operation due to transferring a large amount of data from across machines over the network. When moving data across a network, the data usually goes through a serialization and deserialization process. Imagine performing a join on two large datasets where the size of each one is 100 GB. In this scenario, it moves approximately 200GB of data around. It is not possible to completely avoid a shuffle hash join when joining two large datasets. Still, it is important to be mindful about reducing the frequency of joining them whenever possible.

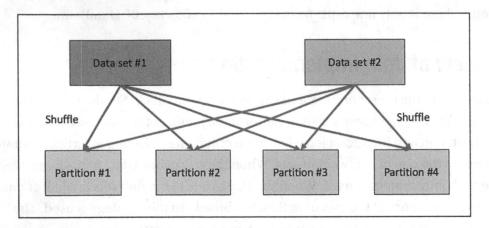

Figure 4-4. *Shuffle hash join*

Broadcast Hash Join

This join strategy is applicable when one of the datasets is small enough to fit into memory. Knowing that the shuffle hash join is an expensive operation, the broadcast hash join avoids shuffling both datasets and shuffles only the smaller one. Like the shuffle hash join strategy, this one also consists of two steps. The first step is to broadcast a copy of the smaller dataset to each of the larger dataset's partitions. The second step is to iterate through each row in the larger dataset and look up the corresponding rows in the smaller dataset with match column values. Figure 4-5 shows the broadcasting of the small dataset.

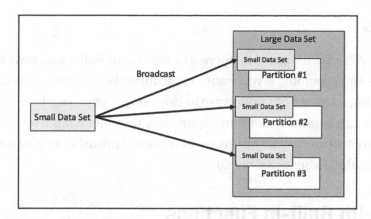

Figure 4-5. *Broadcast hash join*

It is easy to understand that a broadcast hash join is preferred when it is applicable. For the most part, Spark SQL can automatically figure out when to use broadcast hash join or shuffle hash join based on the statistics it has about datasets while reading them. However, it is feasible to provide a hint to Spark SQL to use broadcast hash join when using the join transformation. Listing 4-33 provides an example of doing that.

Listing 4-33. Provide a Hint to Use Broadcast Hash Join

```
import org.apache.spark.sql.functions.broadcast
```

```
// Use broadcast hash join strategy and print out execution plan
employeeDF.join(broadcast(deptDF), employeeDF.col("dept_no") === deptDF.
col("id")).explain()
```

```
// User broadcast hash join hint in a SQL statement
spark.sql("select /*+ MAPJOIN(departments) */ * from employees JOIN
departments on dept_no == id").explain()
```

```
== Physical Plan ==
*BroadcastHashJoin [dept_no#30L], [id#41L], Inner, BuildRight
:- LocalTableScan [first_name#29, dept_no#30L]
+- BroadcastExchange HashedRelationBroadcastMode(List(input[0, bigint,
false]))
   +- LocalTableScan [id#41L, name#42]
```

Functions

The DataFrame APIs are designed to operate or transform individual rows in a dataset, such as filtering and grouping. If you want to transform the column value of each row, such as converting a string from uppercase to camel case, you use a function. Functions are methods that are applied to columns. Spark SQL provides a large set of commonly needed functions and an easy way to create new ones. Approximately 30 new built-in functions were added in Spark 3.0 version.

Working with Built-in Functions

To be effective and productive at using Spark SQL to perform distributed data manipulations, you must be proficient at working with Spark SQL built-in functions. These built-in functions are designed to generate optimized code for execution at runtime, so it is best to take advantage of them before coming up with your own functions. One commonality among these functions is they are designed to take one or more columns of the same row as the input, and they return only a single column as the output. Spark SQL provides more than 200 built-in functions, and they are grouped into different categories. These functions can be used in DataFrame operations, such as `select`, `filter`, and `groupBy`.

For a complete list of built-in functions, refer to the Spark API Scala documentation at `https://spark.apache.org/docs/latest/api/scala/org/apache/spark/sql/functions$.html`. Table 4-3 classifies them into different categories.

Table 4-3. *A Subset of Built-in Functions for Each Category*

Category	Description
Date time	unix_timestamp, from_unixtime, to_date, current_date, current_timesatmp, date_add, date_sub, add_months, datediff, months_between, dayofmonth, dayofyear, weekofyear, second, minute, hour, month, make_date, make_timestamp, make_interval
String	concat, length, levenshtein, locate, lower, upper, ltrim, rtrim, trim, lpad, rpad, repeat, reverse, split, substring, base64
Math	cos, acos, sin, asin, tan, atan, ceil, floor, exp, factorial, log, pow, radian, degree, sqrt, hex, unhex
Cryptography	cr32, hash, md5, sha1, sha2
Aggregation	approx._count_distinct, countDistinct, sumDistinct, avg, corr, count, first, last, max, min, skewness, sum,
Collection	array_contain, explode, from_json, size, sort_array, to_json, size
Window	dense_rank, lag, lead, ntile, rank, row_number
Misc.	coalesce, isNan, isnull, isNotNull, monotonically_increasing_id, lit, when

Most of these functions are easy to understand and straightforward to use. The following sections provide working examples of some of the interesting ones.

Working with Date Time Functions

The more you use Spark to perform data analysis, the more chance you encounter datasets that have one more date or time-related columns. The Spark built-in data time functions broadly fall into the following three categories: converting the date or timestamp from one format to another, performing a data-time calculation, and extracting specific values from a date or timestamp, such as year, month, day of the week, and so on.

The date-time conversion functions help convert a time string into either a date, timestamp, or Unix timestamp and vice versa. Internally, it uses the Java date format pattern syntax, which is documented at `http://docs.oracle.com/javase/tutorial/i18n/format/simpleDateFormat.html`. The default date format these functions use is yyyy-MM-dd HH:mm:ss. Therefore, if your date or timestamp column's date format is different, you need to provide that pattern to these conversion functions. Listing 4-34 shows an example of converting date and timestamp in string type to Spark date and timestamp type.

Listing 4-34. Converting date and timestamp String to Spark Date and Timestamp Type

```
// the last two columns don't follow the default date format
val testDF = Seq((1, "2018-01-01", "2018-01-01 15:04:58:865",
                "01-01-2018", "12-05-2017 45:50"))
              .toDF("id", "date", "timestamp", "date_str",
                  "ts_str")

// convert these strings into date, timestamp and unix timestamp
// and specify a custom date and timestamp format
val testResultDF = testDF.select(to_date('date).as("date1"),
                      to_timestamp('timestamp).as("ts1"),
                  to_date('date_str,"MM-dd-yyyy").as("date2"),
          to_timestamp('ts_str, "MM-dd-yyyy mm:ss").as("ts2"),
          unix_timestamp('timestamp).as("unix_ts"))
                  .show(false)

// date1 and ts1 are of type date and timestamp respectively
testResultDF.printSchema
 |-- date1: date (nullable = true)
 |-- ts1: timestamp (nullable = true)
 |-- date2: date (nullable = true)
 |-- ts2: timestamp (nullable = true)
 |-- unix_ts: long (nullable = true)

testDateResultDF.show
+----------+-------------------+----------+-------------------+-----------+
|    date1|                ts1|    date2|                ts2|   unix_ts|
+----------+-------------------+----------+-------------------+-----------+
|2018-01-01|2018-01-01 15:04:58|2018-01-01|2017-12-05 00:45:50| 1514847898|
+----------+-------------------+----------+-------------------+-----------+
```

It is just as easy to convert a date or timestamp to a time string by using the date_ format function with a custom date format or using the from_unixtime function to convert a Unix timestamp (in seconds) to a time string. Listing 4-35 shows examples of the conversions.

Listing 4-35. Converting Date, Timestamp, and Unix Timestamp to Time String

```
testResultDF.select(date_format('date1,"dd-MM-YYYY").as("date_str"),date_
format('ts1, "dd-MM-YYYY HH:mm:ss").as("ts_str"),
 from_unixtime('unix_ts,"dd-MM-YYYY HH:mm:ss").as("unix_ts_str"))
            .show

+-------------+-----------------------+-----------------------+
|    date_str |                ts_str |           unix_ts_str |
+-------------+-----------------------+-----------------------+
|  01-01-2018 |    01-01-2018 15:04:58 |    01-01-2018 15:04:58 |
+-------------+-----------------------+-----------------------+
```

The date-time calculation functions are useful for figuring out the difference between two dates or timestamps and the ability to perform date or time arithmetic. Listing 4-36 shows working examples of date-time calculation.

Listing 4-36. Date Time Calculation Examples

```
val employeeData = Seq(("John", "2016-01-01", "2017-10-15"),
                       ("May", "2017-02-06", "2017-12-25"))
                   .toDF("name", "join_date", "leave_date")

employeeData.show
+------+---------------+--------------+
| name |     join_date |   leave_date |
+------+---------------+--------------+
| John |    2016-01-01 |   2017-10-15 |
|  May |    2017-02-06 |   2017-12-25 |
+------+---------------+--------------+
```

```
// perform date and month calculations
employeeData.select('name,
            datediff('leave_date, 'join_date).as("days"),
            months_between('leave_date, 'join_date).as("months"),
            last_day('leave_date).as("last_day_of_mon"))
        .show
```

```
+------+------+---------------+----------------------+
| name| days|         months|       last_day_of_mon|
+------+------+---------------+----------------------+
|  John|  653|     21.4516129|            2017-10-31|
|   May|  322|    10.61290323|            2017-12-31|
+------+------+---------------+----------------------+
```

```
// perform date addition and subtraction
val oneDate = Seq(("2018-01-01")).toDF("new_year")
oneDate.select(date_add('new_year, 14).as("mid_month"),
            date_sub('new_year, 1).as("new_year_eve"),
            next_day('new_year, "Mon").as("next_mon"))
        .show
```

```
+--------------+--------------------+----------------+
|     mid_month|        new_year_eve|        next_mon|
+--------------+--------------------+----------------+
|    2018-01-15|          2017-12-31|      2018-01-08|
+--------------+--------------------+----------------+
```

The ability to extract specific fields from a date or timestamp value such as year, month, hour, minutes, and seconds is convenient. For example, when there is a need to group all the stock transactions by quarter, month, or week, you can just extract that information from the transaction date and group by those values. Listing 4-37 shows how easy it is to extract fields out of a date or timestamp.

Listing 4-37. Extract Specific Fields from a Date Value

```
val valentimeDateDF = Seq((("2018-02-14 05:35:55")).toDF("date")

valentimeDateDF.select(year('date).as("year"),
                  quarter('date).as("quarter"),
                  month('date).as("month"),
                  weekofyear('date).as("woy"),
                  dayofmonth('date).as("dom"),
                  dayofyear('date).as("doy"),
                  hour('date).as("hour"),
                  minute('date).as("minute"),
                  second('date).as("second"))
        .show
```

```
+-----+--------+------+-----+-----+-----+------+-------+--------+
| year| quarter| month|  woy|  dom|  doy| hour| minute|  second|
+-----+--------+------+-----+-----+-----+------+-------+--------+
| 2018|       1|     2|    7|   14|   45|    5|     35|      55|
+-----+--------+------+-----+-----+-----+------+-------+--------+
```

Working with String Functions

Undoubtedly most columns in the majority of datasets are of string type. The Spark SQL built-in string functions provide versatile and powerful ways of manipulating this type of column. These functions fall into two broad buckets. The first one is about transforming a string, and the second one is about applying regular expressions either to replace some part of a string or to extract certain parts of a string based on a pattern.

There are many ways to transform a string. The most common ones are trimming, padding, uppercasing, lowercasing, and concatenating. Trimming is about removing the spaces on the left side or right side of a string, or both. Padding is about adding characters to the left side or the right side of a string. Listing 4-38 demonstrates the various ways of transforming a string using the various built-in string functions.

Listing 4-38. Different Ways of Transforming a String With Built-in String Functions

```
val sparkDF = Seq((" Spark ")).toDF("name")

// trimming - removing spaces on the left side, right side of a string, or both
// trim removes spaces on both sides of a string
// ltrim only removes spaces on the left side of a string
// rtrim only removes spaces on the right side of a string
sparkDF.select(trim('name).as("trim"),
               ltrim('name).as("ltrim"),
               rtrim('name).as("rtrim"))
          .show
+-----+----------+---------+
| trim|     ltrim|    rtrim|
+-----+----------+---------+
|Spark|    Spark |    Spark|
+-----+----------+---------+

// padding a string to a specified length with given pad string
// first trim spaces around string "Spark" and then pad it so the final
   length is 8 characters long
// lpad pads the left side of the trim column with - to the length of 8
// rpad pads the right side of the trim colum with = to the length of 8
sparkDF.select(trim('name).as("trim"))
       .select(lpad('trim, 8, "-").as("lpad"),
               rpad('trim, 8, "=").as("rpad"))
       .show

+---------+-------------+
|     lpad|         rpad|
+---------+-------------+
| ---Spark|     Spark===|
+---------+-------------+
```

```
// transform a string with concatenation, uppercase, lowercase and reverse
val sparkAwesomeDF = Seq(("Spark", "is", "awesome"))
                     .toDF("subject", "verb", "adj")

sparkAwesomeDF.select(concat_ws(" ",'subject, 'verb,
                                'adj).as("sentence"))
          .select(lower('sentence).as("lower"),
                  upper('sentence).as("upper"),
                  initcap('sentence).as("initcap"),
                  reverse('sentence).as("reverse"))
          .show
+-----------------+-----------------+-----------------+-----------------+
|            lower|            upper|          initcap|          reverse|
+-----------------+-----------------+-----------------+-----------------+
| spark is awesome| SPARK IS AWESOME| Spark Is Awesome| emosewa si krapS|
+-----------------+-----------------+-----------------+-----------------+

// translate from one character to another
sparkAwesomeDF.select('subject, translate('subject, "ar",
                                    "oc").as("translate"))
          .show
+---------+-----------+
|  subject|  translate|
+---------+-----------+
|    Spark|      Spock|
+---------+-----------+
```

Regular expressions are a powerful and flexible way to replace some portion of a string or extract substrings out of a string. The regexp_extract and regexp_replace functions are designed specifically for those purposes. Spark leverages the Java regular expressions library for the underlying implementation of these two string functions.

The input parameters to the regexp_extract function are a string column, a pattern to match, and a group index. There could be multiple pattern matches in a string; therefore, the group index (starts with 0) is needed to identify which one. If there is no match for the specified pattern, this function returns an empty string. Listing 4-30 is an example of working with the regexp_extract function.

Listing 4-39. Using regexp_extract string Function to Extract "fox" Out Using a Pattern

```
val rhymeDF = Seq(("A fox saw a crow sitting on a tree singing
                    \"Caw! Caw! Caw!\"")).toDF("rhyme")

// using a pattern
rhymeDF.select(regexp_extract('rhyme,"[a-z]*o[xw]",0)
                            .as("substring")).show

+------------+
|   substring|
+------------+
|         fox|
+------------+
```

The input parameters to the regexp_replace string function are the string column, a pattern to match, and a value to replace with. Listing 4-40 is an example of working with the regexp_replace function.

Listing 4-40. Using regexp_replace String Function to Replace "fox" and "crow" with "animal"

```
val rhymeDF = Seq(("A fox saw a crow sitting on a tree singing
                    \"Caw! Caw! Caw!\"")).toDF("rhyme")

// both lines below produce the same output
rhymeDF.select(regexp_replace('rhyme, "fox|crow", "animal")
                            .as("new_rhyme"))
      .show(false)

rhymeDF .select(regexp_replace('rhyme, "[a-z]*o[xw]", "animal")
                .as("new_rhyme"))
        .show(false)

+-------------------------------------------------------------+
|                           new_rhyme                         |
+-------------------------------------------------------------+
|A animal saw a animal sitting on a tree singing "Caw! Caw! Caw!"|
+-------------------------------------------------------------+
```

Working with Math Functions

The second most common column type is numerical type. This is especially true in customer transactions or IoT sensor-related datasets. Most of the math functions are self-explanatory and easy to use. This section covers one useful and commonly used function called round, which performs the half-up rounding of a numeric value based on the given scale. The scale determines the number of decimal points to round up to. There are two variations of this function. The first one takes a column with a floating-point value and a scale, and the second one takes only a column with a floating-point value. The second variation calls the first one with a value of 0 for the scale. Listing 4-41 demonstrates the behavior of the round function.

Listing 4-41. Demonstrates the Behavior of round with Various Scales

```
val numberDF =Seq((3.14159, 3.5, 2018)).toDF("pie","gpa", "year")
numberDF.select(round('pie).as("pie0"),
                round('pie, 1).as("pie1"),
                round('pie, 2).as("pie2"),
                round('gpa).as("gpa"),
                round('year).as("year"))
      .show
// because it is a half-up rounding, the gpa value is rounded up to 4.0
+-----+------+-----+-----+------+
| pie0| pie1| pie2| gpa| year|
+-----+------+-----+-----+------+
| 3.0|   3.1| 3.14| 4.0| 2018|
+-----+------+-----+-----+------+
```

Working with Collection Functions

The collection functions are designed to work with complex data types such as arrays, maps, or structs. This section covers the two specific types of collection functions. The first one is about working with an array data type. The second one is about working with the JSON data format.

Instead of a single scalar value, sometimes a particular column in a dataset contains a list of values. One way to model that is by using an array data type. For example, let say there is a dataset about tasks that need to be performed per day. In this dataset, each row represents a list of tasks per day. Each row consists of two columns. One column contains the date, and the other column contains a list of tasks. You can use the array-related collection functions to easily get the array size, check for the existence of a value, or sort the array. Listing 4-42 contains examples of working with the various array-related functions.

Listing 4-42. Using Array Collection Functions to Manipulate a List of Tasks

```
// create an tasks DataFrame
val tasksDF = Seq(("Monday", Array("Pick Up John",
                                   "Buy Milk", "Pay Bill")))
                  .toDF("day", "tasks")

// schema of tasksDF
tasksDF.printSchema
 |-- day: string (nullable = true)
 |-- tasks: array (nullable = true)
 |    |-- element: string (containsNull = true)

// get the size of the array, sort it, and check to see if a particular
value exists in the array
tasksDF.select('day, size('tasks).as("size"),
               sort_array('tasks).as("sorted_tasks"),
               array_contains('tasks, "Pay Bill").as("payBill"))
       .show(false)
+---------+-----+------------------------------------+---------+
|   day   | size|            sorted_ta               | payBill|
+---------+-----+------------------------------------+---------+
|   Monday|  3  | [Buy Milk, Pay Bill, Pick Up John]|   true  |
+---------+-----+------------------------------------+---------+

// the explode function will create a new row for each element in the array
tasksDF.select('day, explode('tasks)).show
+----------+------------------+
|    day|               col|
+----------+------------------+
|    Monday|      Pick Up John|
|    Monday|          Buy Milk|
|    Monday|          Pay Bill|
+----------+------------------+
```

Many unstructured datasets are in the form of JSON, which is a popular self-describing data format. A common example is to encode the Kafka message payload in JSON format. Since this format is widely supported in most programming languages, a Kafka consumer written in one of these programming languages can easily decode those Kafka messages. The JSON-related collection functions are useful for converting a JSON string to and from a struct data type. The main functions are from_json and to_json. Once a JSON string is converted to a Spark struct data type, you can easily extract those values. Listing 4-43 shows examples of working with from_json and to_json functions.

Listing 4-43. Examples of Using from_json and to_json Functions

```
import org.apache.spark.sql.types._
// create a string that contains JSON string
val todos = """{"day": "Monday","tasks": ["Pick Up John",
              "Buy Milk","Pay Bill"]}"""

val todoStrDF = Seq((todos)).toDF("todos_str")

// at this point, todoStrDF is DataFrame with one column with string data type
todoStrDF.printSchema
 |-- todos_str: string (nullable = true)

// in order to convert a JSON string into a Spark struct data type, we need
to describe its structure to Spark
val todoSchema = new StructType().add("day", StringType)
                       .add("tasks",  ArrayType(StringType))

// use from_json to convert JSON string
val todosDF = todoStrDF.select(from_json('todos_str, todoSchema)
                     .as("todos"))

// todos is a struct data type that contains two fields: day and tasks
todosDF.printSchema
|-- todos: struct (nullable = true)
|     |-- day: string (nullable = true)
|     |-- tasks: array (nullable = true)
|     |     |-- element: string (containsNull = true)
```

```
// retrieving value out of struct data type using the getItem function of
Column class
todosDF.select('todos.getItem("day"), 'todos.getItem("tasks"),
          'todos.getItem("tasks").getItem(0).as("first_task"))
      .show(false)

+-----------+----------------------------------------+------------+
| todos.day| todos.tasks                             | first_task |
+-----------+----------------------------------------+------------+
|  Monday   | [Pick Up John, Buy Milk, Pay Bill]| Pick Up John|
+-----------+----------------------------------------+------------+

// to convert a Spark struct data type to JSON string, we can use to_json
function
todosDF.select(to_json('todos)).show(false)

+----------------------------------------------------------------+
|                      structstojson(todos)                       |
+----------------------------------------------------------------+
|{"day":"Monday","tasks":["Pick Up John","Buy Milk","Pay Bill"]}|
+----------------------------------------------------------------+
```

Working with Miscellaneous Functions

A few of the functions in the miscellaneous category are interesting and can be useful in certain situations. This section covers the following functions: monotonically_increasing_id, when, coalesce, and lit.

Sometimes there is a need to generate monotonically increasing unique, but not consecutive, IDs for each row in the dataset. It is quite an interesting problem if you spend some time thinking about it. For example, if a dataset has 200 million rows and is spread across many partitions (machines), how do you ensure the values are unique and increasing simultaneously? This is the job of the monotonically_increasing_id function, which generates IDs as 64-bit integers. The key part of its algorithm is that it places the partition ID in the upper 31 bits of the generated IDs. Listing 4-44 shows an example of using the monotonically_increasing_id function.

Listing 4-44. monotonically_increasing_id in Action

```
// first generate a DataFrame with values from 1 to 10
// and spread them across 5 partitions
val numDF = spark.range(1,11,1,5)

// verify that there are 5 partitions
numDF.rdd.getNumPartitions
Int = 5

// now generate the monotonically increasing numbers
// and see which ones are in which partition
numDF.select('id, monotonically_increasing_id().as("m_ii"),
           spark_partition_id().as("partition")).show

+----+--------------+-----------+
| id|          m_ii| partition|
+----+--------------+-----------+
|   1|             0|          0|
|   2|             1|          0|
|   3|    8589934592|          1|
|   4|    8589934593|          1|
|   5|   17179869184|          2|
|   6|   17179869185|          2|
|   7|   25769803776|          3|
|   8|   25769803777|          3|
|   9|   34359738368|          4|
|  10|   34359738369|          4|
+----+--------------+-----------+

// the above table shows the values in m_ii columns have a different range
in each partition.
```

If there is a need to evaluate a value against a list of conditions and return a value, then a typical solution is to use a switch statement, which is available in most high-level programming languages. When there is a need to do this with the value of a column in DataFrame, you can use the when function for this use case. Listing 4-45 is an example of using the when function.

Listing 4-45. Use the when Function to Convert a Numeric Value to a String

```
// create a DataFrame with values from 1 to 7 to represent each day of the week
val dayOfWeekDF = spark.range(1,8,1)

// convert each numerical value to a string

dayOfWeekDF.select('id, when('id === 1, "Mon")
                        .when('id === 2, "Tue")
                        .when('id === 3, "Wed")
                        .when('id === 4, "Thu")
                        .when('id === 5, "Fri")
                        .when('id === 6, "Sat")
                        .when('id === 7, "Sun").as("dow"))
        .show

+---+----+
| id| dow|
+---+----+
|  1| Mon|
|  2| Tue|
|  3| Wed|
|  4| Thu|
|  5| Fri|
|  6| Sat|
|  7| Sun|
+---+----+

// to handle the default case when we can use the otherwise function of the
column class
dayOfWeekDF.select('id, when('id === 6, "Weekend")
                        .when('id === 7, "Weekend")
                        .otherwise("Weekday").as("day_type"))
        .show
```

```
+---+--------+
| id|day_type|
+---+--------+
|  1| Weekday|
|  2| Weekday|
|  3| Weekday|
|  4| Weekday|
|  5| Weekday|
|  6| Weekend|
|  7| Weekend|
+------------+
```

When working with data, it is important to handle null values properly. One of the ways to do that is to convert them to some other values that represent null in your data processing logic. Borrowing from the SQL world, Spark provides a coalesce that takes one or more column values and returns the first one that is not null. Each argument in the coalesce must be of type Column, so if you want to fill in a literal value, you can leverage the lit function. This function works because it takes a literal value and returns an instance of the Column class that wraps the input. Listing 4-46 is an example of using both coalesce and lit functions together.

Listing 4-46. Using coalesce to Handle null Value in a Column

```
// create a movie with null title
case class Movie(actor_name:String, movie_title:String,
                 produced_year:Long)

val badMoviesDF = Seq( Movie(null, null, 2018L),
                       Movie("John Doe", "Awesome Movie", 2018L))
                 .toDF

// use coalesce function to handle null value in the title column
badMoviesDF.select(coalesce('actor_name,
                   lit("no_name")).as("new_title"))
     .show
```

```
+-------------+
|   new_title|
+-------------+
|     no_name|
|    John Doe|
+-------------+
```

Working with User-Defined Functions (UDFs)

Even though Spark SQL provides a large set of built-in functions for most common use cases, there are always cases where none of those functions can provide the functionality your use cases need. However, don't despair. Spark SQL provides a simple facility to write user-defined functions (UDFs) and uses them in your Spark data processing logic or applications similarly to using built-in functions. UDFs are effectively one of the ways you can extend Spark's functionality to meet your specific needs.

Another thing that I like about Spark because UDFs can be written in either Python, Java, or Scala, and they can leverage and integrate with any necessary libraries. Since you can use a programming language that you are most comfortable with to write UDFs, it is extremely easy and fast to develop and test UDFs.

Conceptually, UDFs are just regular functions that take some inputs and provide an output. Although UDFs can be written in either Scala, Java, or Python, you must be aware of the performance differences when UDFs are written in Python. UDFs must be registered with Spark before they are used, so Spark knows to ship them to executors to be used and executed. Given that executors are JVM processes written in Scala, they can execute Scala or Java UDFs natively in the same process. If a UDF is written in Python, then an executor can't execute it natively, and therefore it must spawn a separate Python process to execute the Python UDF. In addition to the cost of spawning a Python process, there is a high cost in terms of serializing data back and forth for each row in the dataset.

There are three steps involved in working with UDFs. The first one is to write a function and test it. The second step is to register that function with Spark by passing in the function name and its signature to Spark's udf function. The last step is to use UDF in either the DataFrame code or when issuing SQL queries. The registration process is slightly different when using a UDF within SQL queries. Listing 4-47 demonstrates the three steps with a simple UDF.

Listing 4-47. A Simple UDF in Scala to Convert Numeric Grades to Letter Grades

```scala
import org.apache.spark.sql.functions.udf

// create student grades DataFrame
case class Student(name:String, score:Int)

val studentDF = Seq(Student("Joe", 85),  Student("Jane",
90),  Student("Mary", 55)).toDF()

// register as a view
studentDF.createOrReplaceTempView("students")

// create a function to convert grade to a letter grade
def letterGrade(score:Int) : String = {
  score match {
    case score if score > 100 => "Cheating"
    case score if score >= 90 => "A"
    case score if score >= 80 => "B"
    case score if score >= 70 => "C"
    case _ => "F"
  }
}

// register as an UDF
val letterGradeUDF = udf(letterGrade(_:Int):String)

// use the UDF to convert scores to letter grades
studentDF.select($"name",$"score",
                letterGradeUDF($"score").as("grade")).show

+----+-----+-----+
|name|score|grade|
+----+-----+-----+
| Joe|   85|    B|
|Jane|   90|    A|
|Mary|   55|    F|
+----+-----+-----+
```

```
// register as UDF to use in SQL
spark.sqlContext.udf.register("letterGrade",
                              letterGrade(_: Int): String)

spark.sql("select name, score, letterGrade(score) as grade from students").
show
```

```
+----+-----+-----+
|name|score|grade|
+----+-----+-----+
| Joe|   85|    B|
|Jane|   90|    A|
|Mary|   55|    F|
+----+-----+-----+
```

Advanced Analytics Functions

The previous sections covered the built-in functions Spark SQL provides for basic analytic needs such as aggregation, joining, pivoting, and grouping. All those functions take one or more values from a single row and produce an output value, or they take a group of rows and return an output.

This section covers the advanced analytics capabilities Spark SQL offers. The first one is about multidimensional aggregations, which is useful for use cases involving hierarchical data analysis. Calculating subtotals and totals across a set of grouping columns is commonly needed. The second capability is about performing aggregations based on time windows, which is useful when working with time-series data such as transactions or sensor values from IoT devices. The third one is the ability to perform aggregations within a logical grouping of rows, referred to as a window. This capability enables you to easily perform calculations such as a moving average, a cumulative sum, or the rank of each row.

Aggregation with Rollups and Cubes

Rollups and cube are more advanced versions of grouping on multiple columns, and they generate subtotals and grand totals across the combinations and permutations of those columns. The order of the provided set of columns is treated as a hierarchy for grouping.

Rollups

When working with hierarchical data such as the revenue data that spans different departments and divisions, rollups can easily calculate the subtotals and a total across them. Rollups respect the given hierarchy of the given set of rollup columns and always start the rolling up process with the first column in the hierarchy. The total is listed in the output, where all the column values are null. Listing 4-48 demonstrates how a rollup works.

Listing 4-48. Performing Rollups with Flight Summary Data

```
// read in the flight summary data
val flight_summary = spark.read.format("csv")
                        .option("header", "true")
                        .option("inferSchema","true")
        .load(<path>/chapter4/data/     flights/flight-summary.csv)

// filter data down to smaller size to make it easier to see the rollups
result
val twoStatesSummary = flight_summary.select('origin_state,
                                    'origin_city,'count)
    .where('origin_state === "CA" || 'origin_state === "NY")
    .where('count > 1 && 'count < 20)
    .where('origin_city =!= "White Plains")
    .where('origin_city =!= "Newburgh")
    .where('origin_city =!= "Mammoth Lakes")
    .where('origin_city =!= "Ontario")

// let's see what the data looks like
twoStatesSummary.orderBy('origin_state).show
```

origin_state	origin_city	count
CA	San Diego	18
CA	San Francisco	5
CA	San Francisco	14
CA	San Diego	4

```
|          CA| San Francisco|     2|
|          NY|      New York|     4|
|          NY|      New York|     2|
|          NY|        Elmira|    15|
|          NY|        Albany|     5|
|          NY|        Albany|     3|
|          NY|      New York|     4|
|          NY|        Albany|     9|
|          NY|      New York|    10|
+------------+--------------+------+
```

```
// perform the rollup by state, city,
// then calculate the sum of the count,and finally order by null last
twoStatesSummary.rollup('origin_state, 'origin_city)
             .agg(sum("count") as "total")
             .orderBy('origin_state.asc_nulls_last,
                      'origin_city.asc_nulls_last)
             .show
```

```
+------------+--------------+------+
| origin_state|   origin_city| total|
+------------+--------------+------+
|          CA|     San Diego|    22|
|          CA| San Francisco|    21|
|          CA|          null|    43|
|          NY|        Albany|    17|
|          NY|        Elmira|    15|
|          NY|      New York|    20|
|          NY|          null|    52|
|        null|          null|    95|
+------------+--------------+------+
```

This output shows the subtotals per state on the third and seventh lines. The last line shows the total with a null value in both the original_state and origin_city columns. The trick is to sort with the asc_nulls_last option, so Spark SQL order null values last.

Cubes

A cube is a more advanced version of a rollup. It performs the aggregations across all the combinations of the grouping columns. Therefore, the result includes what a rollup provides, as well as other combinations. In the cubing by origin_state and origin_city example, the result includes the aggregation for each of the original cities. The way to use the cube function is similar to how you use the rollup function.

Listing 4-49 is an example.

Listing 4-49. Performing a Cube Across the origin_state and origin_city Columns

```
// perform the cube across origin_state and origin_city
twoStatesSummary.cube('origin_state, 'origin_city)
        .agg(sum("count") as "total")
        .orderBy('origin_state.asc_nulls_last,
                'origin_city.asc_nulls_last)
        .show

+------------+-------------+-----+
|origin_state|  origin_city|total|
+------------+-------------+-----+
|          CA|    San Diego|   22|
|          CA|San Francisco|   21|
|          CA|         null|   43|
|          NY|       Albany|   17|
|          NY|       Elmira|   15|
|          NY|     New York|   20|
|          NY|         null|   52|
|        null|       Albany|   17|
|        null|       Elmira|   15|
|        null|     New York|   20|
|        null|    San Diego|   22|
|        null|San Francisco|   21|
|        null|         null|   95|
+------------+-------------+-----+
```

In the table, the lines with a null value in the origin_state column represent an aggregation of all the cities in a state. Therefore, the result of a cube always has more rows than the result of a rollup.

Aggregation with Time Windows

Aggregation with time windows was introduced in Spark 2.0 to make it easy to work with time-series data, consisting of a series of data points in time order. This kind of dataset is common in industries such as finance or telecommunications. For example, the stock market transaction dataset has the transaction date, opening price, close price, volume, and other pieces of information for each stock symbol. Time window aggregations can help answer questions such as the weekly average closing price of Apple stock or the monthly moving average closing price of Apple stock across each week.

Window functions come in a few versions, but they all require a timestamp type column and a window length, specified in seconds, minutes, hours, days, or weeks. The window length represents a time window with a start time and end time, and it determines which bucket a particular piece of time-series data should belong to. Another version takes additional input for the sliding window size, which tells how much a time window should slide when calculating the next bucket. These versions of the window function are the implementations of the tumbling window and sliding window concepts in world event processing, and they are described in more detail in Chapter 6.

The following examples use the Apple stock transactions, which can be found on the Yahoo! Finance website at `https://in.finance.yahoo.com/q/hp?s=AAPL`. Listing 4-50 calculates the weekly average price of Apple stock based on one year of data.

Listing 4-50. Using the Time Window Function to Calculate the Average Closing Price of Apple Stock

```
val appleOneYearDF = spark.read.format("csv")
                        .option("header", "true")
                        .option("inferSchema","true")
            .load("<path>/chapter5/data/stocks/aapl-2017.csv")

// display the schema, the first column is the transaction date
appleOneYearDF.printSchema
 |-- Date: string (nullable = true)
 |-- Open: double (nullable = true)
```

```
|-- High: double (nullable = true)
|-- Low: double (nullable = true)
|-- Close: double (nullable = true)
|-- Adj Close: double (nullable = true)
|-- Volume: integer (nullable = true)
```

```scala
// calculate the weekly average price using window function inside the
groupBy transformation
// this is an example of the tumbling window, aka fixed window
val appleWeeklyAvgDF = appleOneYearDF.
        groupBy(window('Date, "1 week"))
        .agg(avg("Close"). as("weekly_avg"))
```

```scala
// the result schema has the window start and end time
appleWeeklyAvgDF.printSchema
 |-- window: struct (nullable = false)
 |    |-- start: timestamp (nullable = true)
 |    |-- end: timestamp (nullable = true)
 |-- weekly_avg: double (nullable = true)
```

```scala
// display the result with ordering by start time and
// round up to 2 decimal points
appleWeeklyAvgDF.orderBy("window.start")
            .selectExpr("window.start",
                    "window.end","round(weekly_avg, 2) as
                    weekly_avg")
            .show(5)
```

```
// notice the start time is inclusive and end time is exclusive
+-------------------+-------------------+--------------+
|              start|                end|    weekly_avg|
+-------------------+-------------------+--------------+
| 2016-12-28 16:00:00| 2017-01-04 16:00:00|         116.08|
| 2017-01-04 16:00:00| 2017-01-11 16:00:00|         118.47|
| 2017-01-11 16:00:00| 2017-01-18 16:00:00|         119.57|
| 2017-01-18 16:00:00| 2017-01-25 16:00:00|         120.34|
| 2017-01-25 16:00:00| 2017-02-01 16:00:00|         123.12|
+-------------------+-------------------+--------------+
```

Listing 4-50 uses a one-week tumbling window, where there is no overlap.

Therefore, each transaction is used only once to calculate the moving average. The example in Listing 4-51 uses the sliding window. This means some transactions are used more than once in calculating the average monthly moving average. The window size is four weeks, and it slides by one week at a time in each window.

Listing 4-51. Use the Time Window Function to Calculate the Monthly Average Closing Price of Apple Stock

```
// 4 weeks window length and slide by one week each time
val appleMonthlyAvgDF = appleOneYearDF.groupBy(
                    window('Date, "4 week", "1 week"))
                .agg(avg("Close").as("monthly_avg"))

// display the results with order by start time
appleMonthlyAvgDF.orderBy("window.start")
            .selectExpr("window.start", "window.end",
                    "round(monthly_avg, 2) as monthly_avg")
            .show(5)
```

```
+-------------------+-------------------+-----------+
|              start|                end|monthly_avg|
+-------------------+-------------------+-----------+
| 2016-12-07 16:00:00| 2017-01-04 16:00:00|     116.08|
| 2016-12-14 16:00:00| 2017-01-11 16:00:00|     117.79|
| 2016-12-21 16:00:00| 2017-01-18 16:00:00|     118.44|
| 2016-12-28 16:00:00| 2017-01-25 16:00:00|     119.03|
| 2017-01-04 16:00:00| 2017-02-01 16:00:00|     120.42|
+-------------------+-------------------+-----------+
```

Since the sliding window interval is one week, the previous result table shows that the start time difference between two consecutive rows is one week apart. Between two consecutive rows, there are about three weeks of overlapping transactions, which means a transaction is used more than once to calculate the moving average.

Window Functions

You know how to use functions such as concat or round to compute an output from one or more column values of a single row and leverage aggregation functions such as max or sum to compute an output for each group of rows. Sometimes there is a need to operate on a group of rows and return a value for every input row. Window functions provide this unique capability to make it easy to perform calculations such as a moving average, a cumulative sum, or the rank of each row.

There are two main steps for working with window functions. The first one is to define a window specification that defines a logical grouping of rows called a frame, which is the context in which each row is evaluated. The second step is to apply a window function appropriate for the problem you are trying to solve. You learn more about the available window functions in the following sections.

The window specification defines three important components the window functions use. The first component is called partition by, and this is where you specify one or more columns to group the rows by. The second component is called order by, and it defines how the rows should be ordered based on one or more columns and whether the ordering should be in ascending or descending order. Out of the three components, the last one is more complicated and requires a detailed explanation. The last component is called a *frame*, and it defines the boundary of the window in the current row. In other words, the "frame" restricts which rows to be included when calculating a value for the current row. A range of rows to include in a window frame can be specified using the row index or the actual value of the order by expression. The last component is applicable for some of the window functions, and therefore it may not be necessary for some scenarios. A window specification is built using the functions defined in the org.apache.spark.sql.expressions.Window class. The rowsBetween and rangeBetweeen functions define the range by row index and actual value, respectively.

Window functions can be categorized into three different types: ranking functions, analytic functions, and aggregate functions. The ranking and analytic functions are described in Table 4-4 and Table 4-5, respectively. For aggregate functions, you can use any of the aggregation functions as a window function. You can find a complete list of the window functions at https://spark.apache.org/docs/latest/api/java/org/apache/spark/sql/functions.html.

Table 4-4. *Ranking Functions*

Name	Description
rank	Returns the rank or order of rows within a frame based on some sorting order.
dense_rank	Similar to rank, but leaves no gaps in the ranks when there are ties.
percen_rank	Returns the relative rank of rows within a frame.
ntile(n)	Returns the ntile group ID in an ordered window partition. For example, if n is 4, the first quarter of the rows get a value of 1, the second quarter of rows get a value of 2, and so on.
row_number	Returns a sequential number starting with 1 with a frame.

Table 4-5. *Analytic Functions*

Name	Description
cume_dist	Returns the cumulative distribution of values with a frame. In other words, the fraction of rows that are below the current row.
lag(col, offset)	Returns the value of the column that is offset rows before the current row.
lead(col, offset)	Returns the value of the column that is offset rows after the current row.

Let's put the steps together by working through a small sample dataset to demonstrate window function capabilities. Table 4-6 contains the shopping transaction data of two fictitious users: John and Mary.

Table 4-6. *User Shopping Transactions*

Name	Date	Amount
John	2017-07-02	13.35
John	2016-07-06	27.33
John	2016-07-04	21.72
Mary	2017-07-07	69.74
Mary	2017-07-01	59.44
Mary	2017-07-05	80.14

With this shopping transaction data, let's try using window functions to answer the following questions.

- For each user, what are the two highest transaction amounts?

- What is the difference between the transaction amount of each user and their highest transaction amount?

- What is the moving average transaction amount of each user?

- What is the cumulative sum of the transaction amount of each user?

To answer the first question, you apply the rank window function over a window specification that partitions the data by user and sorts it by the amount in descending order. The ranking window function assigns a rank to each row based on the sorting order of each row in each frame. Listing 4-52 is the actual code to solve the first question.

Listing 4-52. Apply the Rank Window Function to Find out the Top Two Transactions per User

```
// small shopping transaction dataset for two users
val txDataDF= Seq(("John", "2017-07-02", 13.35),
                  ("John", "2017-07-06", 27.33),
                  ("John", "2017-07-04", 21.72),
                  ("Mary", "2017-07-07", 69.74),
                  ("Mary", "2017-07-01", 59.44),
                  ("Mary", "2017-07-05", 80.14))
             .toDF("name", "tx_date", "amount")
// import the Window class
import org.apache.spark.sql.expressions.Window

// define window specification to partition by name
// and order by amount in descending amount
val forRankingWindow =
    Window.partitionBy("name").orderBy(desc("amount"))

// add a new column to contain the rank of each row,
// apply the rank function to rank each row
val txDataWithRankDF =
    txDataDF.withColumn("rank", rank().over(forRankingWindow))
```

```
// filter the rows down based on the rank to find
// the top 2 and display the result
txDataWithRankDF.where('rank < 3).show(10)
+------+----------+-------+-----+
| name|   tx_date| amount| rank|
+------+----------+-------+-----+
|  Mary| 2017-07-05|  80.14|    1|
|  Mary| 2017-07-07|  69.74|    2|
|  John| 2017-07-06|  27.33|    1|
|  John| 2017-07-04|  21.72|    2|
+------+----------+-------+-----+
```

The approach for solving the second question involves applying the max function over the amount column across all the partition rows. In addition to partitioning by the username, it needs to define a frame boundary that includes all the rows in each partition. To do that, you use the Window.rangeBetween function with Window.unboundedPreceding as the start value and Window.unboundedFollowing as the end value. Listing 4-53 defines a window specification according to the logic defined earlier and applies the max function over it.

Listing 4-53. Applying the max Window Function to Find the Difference of Each Row and the Highest Amount

```
// use rangeBetween to define the frame boundary that includes
// all the rows in each frame
val forEntireRangeWindow =
    Window.partitionBy("name").orderBy(desc("amount"))
        .rangeBetween(Window.unboundedPreceding,
                    Window.unboundedFollowing)

// apply the max function over the amount column and then compute // the
difference
val amountDifference =
        max(txDataDF("amount")).over(forEntireRangeWindow) -
                                txDataDF("amount")
```

```
// add the amount_diff column using the logic defined above
val txDiffWithHighestDF =
  txDataDF.withColumn("amount_diff", round(amountDifference, 3))

// display the result
txDiffWithHighestDF.show
```

```
+------+----------+-------+------------+
|  name|   tx_date| amount| amount_diff|
+------+----------+-------+------------+
|  Mary| 2017-07-05|  80.14|         0.0|
|  Mary| 2017-07-07|  69.74|        10.4|
|  Mary| 2017-07-01|  59.44|        20.7|
|  John| 2017-07-06|  27.33|         0.0|
|  John| 2017-07-04|  21.72|        5.61|
|  John| 2017-07-02|  13.35|       13.98|
+------+----------+-------+------------+
```

To compute the transaction amount moving average of each user in the order of transaction date, you leverage the avg function to calculate the average amount for each row based on a set of rows in a frame. For this example, you want each frame to include three rows: the current row plus one row before it and one row after it. Depending on a particular use case, the frame might include more rows before and after the current row. Like the previous examples, the window specification partition the data by user, but the rows in each frame are sorted by transaction date. Listing 4-54 shows how to apply the avg function over the window specification described earlier.

Listing 4-54. Applying the Average Window Function to Compute the Moving Average Transaction Amount

```
// define the window specification
// a good practice is to specify the offset relative to
// Window.currentRow

val forMovingAvgWindow =
        Window.partitionBy("name").orderBy("tx_date")
            .rowsBetween(Window.currentRow-1,Window.currentRow+1)
```

```
// apply the average function over the amount column over the
// window specification
// also round the moving average amount to 2 decimals

val txMovingAvgDF = txDataDF.withColumn("moving_avg",
        round(avg("amount").over(forMovingAvgWindow), 2))

// display the result
txMovingAvgDF.show

+------+----------+-------+-----------+
|  name|   tx_date| amount| moving_avg|
+------+----------+-------+-----------+
|  Mary| 2017-07-01|  59.44|      69.79|
|  Mary| 2017-07-05|  80.14|      69.77|
|  Mary| 2017-07-07|  69.74|      74.94|
|  John| 2017-07-02|  13.35|      17.54|
|  John| 2017-07-04|  21.72|       20.8|
|  John| 2017-07-06|  27.33|      24.53|
+------+----------+-------+-----------+
```

To compute the cumulative sum of the transaction amount for each user, apply the sum function over a frame that consists of all the rows up to the current row. The partition by and order by clauses are the same as the moving average example. Listing 4-55 shows how to apply the sum function over the window specification described earlier.

Listing 4-55. Applying the sum Window Function to Compute the Cumulative Sum of Transaction Amount

```
// define the window specification with each frame includes all
// the previous rows and the current row
val forCumulativeSumWindow =
        Window.partitionBy("name").orderBy("tx_date")
            .rowsBetween(Window.unbounded
                        Preceding,Window.currentRow)
```

```
// apply the sum function over the window specification
val txCumulativeSumDF =
    txDataDF.withColumn("culm_sum",round(sum("amount")
            .over(forCumulativeSumWindow),2))

// display the result

txCumulativeSumDF.show
+------+----------+-------+---------+
|  name|   tx_date| amount| culm_sum|
+------+----------+-------+---------+
|  Mary| 2017-07-01|  59.44|    59.44|
|  Mary| 2017-07-05|  80.14|   139.58|
|  Mary| 2017-07-07|  69.74|   209.32|
|  John| 2017-07-02|  13.35|    13.35|
|  John| 2017-07-04|  21.72|    35.07|
|  John| 2017-07-06|  27.33|     62.4|
+------+----------+-------+---------+
```

The default frame of a window specification includes all the preceding rows and up to the current row. In Listing 4-55, it is unnecessary to specify the frame, so you should get the same result. The window function examples were written using the DataFrame APIs. You can achieve the same goals using SQL with the PARTITION BY, ORDER BY, ROWS BETWEEN, and RANGE BETWEEN keywords.

The frame boundary can be specified using the following keywords: UNBOUNDED PRECEDING, UNBOUNDED FOLLOWING, CURRENT ROW, <value> PRECEDING, and <value> FOLLOWING. Listing 4-56 shows examples of using the window functions with SQL.

Listing 4-56. Example of a Window Function in SQL

```
// register the txDataDF as a temporary view called tx_data
txDataDF.createOrReplaceTempView("tx_data")

// use RANK window function to find top two highest transaction amount

spark.sql("select name, tx_date, amount, rank from
(
  select name, tx_date, amount,
```

```
        RANK() OVER (PARTITION BY name ORDER BY amount DESC) as rank
        from tx_data
) where rank < 3").show
```

```
// difference between maximum transaction amount
spark.sql("select name, tx_date, amount, round((max_amount -
            amount),2) as amount_diff from
(
  select name, tx_date, amount, MAX(amount) OVER
   (PARTITION BY name ORDER BY amount DESC
   RANGE BETWEEN UNBOUNDED PRECEDING AND UNBOUNDED FOLLOWING
) as max_amount from tx_data)").show
// moving average
spark.sql("select name, tx_date, amount, round(moving_avg,2) as moving_avg
from
(
    select name, tx_date, amount, AVG(amount) OVER
       (PARTITION BY name ORDER BY tx_date
       ROWS BETWEEN 1 PRECEDING AND 1 FOLLOWING
       ) as moving_avg from tx_data)"
).show
```

```
// cumulative sum
spark.sql("select name, tx_date, amount, round(culm_sum,2) as moving_avg
from
(
    select name, tx_date, amount, SUM(amount) OVER
      (PARTITION BY name ORDER BY tx_date
      ROWS BETWEEN UNBOUNDED PRECEDING AND CURRENT ROW
      ) as culm_sum from tx_data)"
).show
```

When using the window functions in SQL, the partition by, order by, and frame window must be specified in a single statement.

Exploring Catalyst Optimizer

The easiest way to write efficient data processing applications is to not worry about it and automatically optimize your data processing applications. That is the promise of the Spark Catalyst, which is a query optimizer and is the second major component in the Spark SQL module. It plays a major role in ensuring the data processing logic written in either DataFrame APIs or SQL runs efficiently and quickly. It was designed to minimize end-to-end query response times and be extensible such that Spark users can inject user code into the optimizer to perform custom optimization.

At a high level, the Spark Catalyst translates the user-written data processing logic into a logical plan, then optimizes it using heuristics, and finally converts the logical plan to a physical plan. The final step is to generate code based on the physical plan. Figure 4-6 provides a visual representation of the steps.

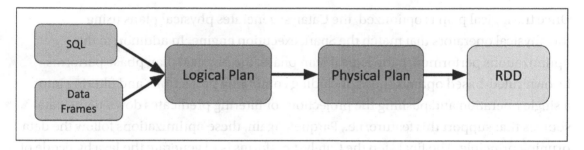

Figure 4-6. *Catalyst optimizer*

Logical Plan

The first step in the Catalyst optimization process is to create a logical plan from either a DataFrame object or the abstract syntax tree of the parsed SQL query. The logical plan is an internal representation of the user data processing logic in a tree of operators and expressions. Next, the Catalyst analyzes the logical plan to resolve references to ensure they are valid. Then it applies a set of rule-based and cost-based optimizations to the logical plan. Both types of optimizations follow the principle of pruning unnecessary data as early as possible and minimizing per-operator cost.

The rule-based optimizations include constant folding, project pruning, predicate pushdown, and others. For example, during this optimization phase, the Catalyst may decide to move the filter condition before performing a join. For curious minds, the list of rule-based optimizations is defined in the `org.apache.spark.sql.catalyst.optimizer.Optimizer` class.

The cost-based optimizations were introduced in Spark 2.2 to enable Catalyst to be more intelligent in selecting the right kind of join based on the statistics of the data being processed. The cost-based optimization relies on the detailed statistics of the columns participating in the filter or join conditions, and that's why the statistics collection framework was introduced. Examples of the statistics include cardinality, the number of distinct values, max/min, and average/max length.

Physical Plan

Once the logical plan is optimized, the Catalyst generates physical plans using the physical operators that match the Spark execution engine. In addition to the optimizations performed in the logical plan phase, the physical plan phase performs its own ruled-based optimizations, including combining projections and filtering into a single operation and pushing the projections or filtering predicates down to the data sources that support this feature, i.e., Parquet. Again, these optimizations follow the data pruning principle. The final step the Catalyst performs is to generate the Java bytecode of the cheapest physical plan.

Catalyst in Action

This section shows how to use the `explain` function of the `DataFrame` class to display the logical and physical plans.

You can call the explain function with the extended argument as a boolean true value to see both the logical and physical plan. Otherwise, this function displays only the physical plan.

The small and somewhat silly example first reads the movie data in Parquet format, performs filtering based on produced_year, adds a column called produced_ decade, and projects the movie_title and produced_decade columns and finally filters rows based on `produced_decade`. The goal here is to prove that the Catalyst performs the predicate pushdown and filtering condition optimizations by examining the generated logical and physical plan by passing a boolean true value to the explain function. In the output, you see four sections: parsed logical plan, analyzed logical plan, optimized logical plan, and physical plan. Listing 4-57 shows how to generate logical and physical plans.

Listing 4-57. Using the explain Function to Generate the Logical and Physical Plans

```
// read movies data in Parquet format
val moviesDF =
    spark.read.load("<path>/book/chapter4/data/movies/movies.
                     parquet")
// perform two filtering conditions
val newMoviesDF = moviesDF.filter('produced_year > 1970)
                          .withColumn("produced_decade",
                          'produced_year + 'produced_year % 10)

val latestMoviesDF = newMoviesDF.select('movie_title,
                                        'produced_decade)
                                .where('produced_decade > 2010)

// display both logical and physical plans
latestMoviesDF.explain(true)

== Parsed Logical Plan ==
'Filter ('produced_decade > 2010)
+- Project [movie_title#673, produced_decade#678L]
   +- Project [actor_name#672, movie_title#673, produced_year#674L,
(produced_year#674L + (produced_year#674L % cast(10 as bigint))) AS
produced_decade#678L]
      +- Filter (produced_year#674L > cast(1970 as bigint))
      +- Relation[actor_name#672,movie_title#673,produced_year#674L] parquet

== Analyzed Logical Plan ==
movie_title: string, produced_decade: bigint
Filter (produced_decade#678L > cast(2010 as bigint))
+- Project [movie_title#673, produced_decade#678L]
```

```
    +- Project [actor_name#672, movie_title#673, produced_year#674L,
(produced_year#674L + (produced_year#674L % cast(10 as bigint))) AS
produced_decade#678L]
        +- Filter (produced_year#674L > cast(1970 as bigint))
        +- Relation[actor_name#672,movie_title#673,produced_year#674L] parquet

== Optimized Logical Plan ==
Project [movie_title#673, (produced_year#674L + (produced_year#674L % 10))
AS produced_decade#678L]
+- Filter ((isnotnull(produced_year#674L) AND (produced_year#674L > 1970))
AND ((produced_year#674L + (produced_year#674L % 10)) > 2010))
    +- Relation[actor_name#672,movie_title#673,produced_year#674L] parquet

== Physical Plan ==
*(1) Project [movie_title#673, (produced_year#674L + (produced_year#674L %
10)) AS produced_decade#678L]
+- *(1) Filter ((isnotnull(produced_year#674L) AND (produced_year#674L >
1970)) AND ((produced_year#674L + (produced_year#674L % 10)) > 2010))
    +- *(1) ColumnarToRow
    +- FileScan parquet [movie_title#673,produced_year#674L] Batched:
true, DataFilters: [isnotnull(produced_year#674L), (produced_year#674L >
1970), ((produced_year#674L + (produced_yea..., Format: Parquet, Location:
InMemoryFileIndex[file:<path>/chapter4/data/movies/..., PartitionFilters:
[], PushedFilters: [IsNotNull(produced_year), GreaterThan(produced_
year,1970)], ReadSchema: struct<movie_title:string,produced_year:bigint>
```

If you carefully analyze the optimized logical plan, you see that it combines both filtering conditions into a single filter. The physical plan shows that Catalyst both pushes down the filtering of produced_year and performs the projection pruning in the FileScan step to optimally read in only the needed data.

In Spark 3.0, a new variation of the explain function was introduced. It takes an input in the form of a string to allow you to specify which of the five modes to see in the output (see Table 4-7).

Table 4-7. *The Various Modes of the Output Format*

Mode	Description
simple	Print only a physical plan.
extended	Print both logical and physical plans.
codegen	Print a physical plan and the generated codes (if they are available).
cost	Print a logical plan and statistics if they are available.
formatted	Split the explain output into two sections; a physical plan outline and details.

The last three options generate new information. It is fascinating to examine the generated Scala code and leave that as an exercise for you. The output of the formatted option is much more readable and easier to understand. Listing 4-58 shows how to use the explain function with the formatted mode.

Listing 4-58. Using the explain Function with formatted Mode

```
latestMoviesDF.explain("formatted")

== Physical Plan ==
* Project (4)
+- * Filter (3)
   +- * ColumnarToRow (2)
   +- Scan parquet  (1)


(1) Scan parquet
Output [2]: [movie_title#673, produced_year#674L]
Batched: true
Location: InMemoryFileIndex [file:<path>/chapter4/data/movies/movies.
parquet]
PushedFilters: [IsNotNull(produced_year), GreaterThan(produced_year,1970)]
ReadSchema: struct<movie_title:string,produced_year:bigint>

(2) ColumnarToRow [codegen id : 1]
Input [2]: [movie_title#673, produced_year#674L]

(3) Filter [codegen id : 1]
Input [2]: [movie_title#673, produced_year#674L]
```

```
Condition : ((isnotnull(produced_year#674L) AND (produced_year#674L >
1970)) AND ((produced_year#674L + (produced_year#674L % 10)) > 2010))

(4) Project [codegen id : 1]
Output [2]: [movie_title#673, (produced_year#674L + (produced_year#674L %
10)) AS produced_decade#678L]
Input [2]: [movie_title#673, produced_year#674L]
```

The output clearly shows Spark's four steps to compute the latestMoviesDF: scan or read the input parquet file, convert the data in columnar format into rows, filter them based on the two specified conditions, and finally project the title and produced decade columns.

Project Tungsten

Starting in 2015, the Spark designers observed that the Spark workloads were increasingly bottlenecked by CPU and memory rather than I/O and network communication. It is a bit counterintuitive but not too surprising, given the advancements on the hardware side like 10Gbps network links and high-speed SSD. Project Tungsten was created to improve the efficiency of using memory and CPU in Spark applications and push the performance closer to the limits of modern hardware. There are three initiatives in the Tungsten project.

- Manage memory explicitly by using off-heap management techniques to eliminate the overhead of the JVM object model and minimize garbage collection.

- Use intelligent cache-aware algorithms and data structures to exploit memory hierarchy.

- Use whole-stage code generation to minimize virtual function calls by combining multiple operators into a single function.

The hard and interesting work that went into the Tungsten project has dramatically improved the Spark execution engine since Spark 2.0. Much of the work in the Tungsten project happens behind the scenes in the execution engine. The following example demonstrates a small glimpse into the whole-stage code generation initiative by examining the physical plan. In the following output, whenever an asterisk (*) appears before an operator, it means the whole-stage code generation is enabled. Listing 4-59 displays the physical plan of filtering and summing integers in a DataFrame.

Listing 4-59. Demonstrating the Whole-Stage Code Generation by Looking at the Physical Plan

```
spark.range(1000).filter("id > 100")
                .selectExpr("sum(id)").explain("formatted")

== Physical Plan ==
* HashAggregate (5)
+- Exchange (4)
   +- * HashAggregate (3)
      +- * Filter (2)
         +- * Range (1)


(1) Range [codegen id : 1]
Output [1]: [id#719L]
Arguments: Range (0, 1000, step=1, splits=Some(12))

(2) Filter [codegen id : 1]
Input [1]: [id#719L]
Condition : (id#719L > 100)

(3) HashAggregate [codegen id : 1]
Input [1]: [id#719L]
Keys: []
Functions [1]: [partial_sum(id#719L)]
Aggregate Attributes [1]: [sum#726L]
Results [1]: [sum#727L]

(4) Exchange
Input [1]: [sum#727L]
Arguments: SinglePartition, ENSURE_REQUIREMENTS, [id=#307]

(5) HashAggregate [codegen id : 2]
Input [1]: [sum#727L]
Keys: []
Functions [1]: [sum(id#719L)]
Aggregate Attributes [1]: [sum(id#719L)#723L]
Results [1]: [sum(id#719L)#723L AS sum(id)#724L]
```

The whole-stage code generation combines the logic of filtering and summing integers into a single function.

Summary

This chapter covered a lot of useful and powerful features available in the Spark SQL module.

- Aggregation is one of the most commonly used features in the world of big data analytics. Spark SQL provides many of the commonly needed aggregation functions such as sum, count, and avg. Aggregation with pivoting provides a nice way of summarizing the data as well as transposing columns into rows.

- Performing any complex and meaningful data analytics or processing often requires joining two or more datasets. Spark SQL supports many of the standard join types that exist in the SQL world.

- Spark SQL comes with a rich set of built-in functions, which should cover most of the common needs for working with strings, math, date and time, and so on. If none meets a particular need of a use case, then it is easy to write a user-defined function that can be used with the DataFrame APIs and SQL queries.

- Window functions are powerful and advanced analytics functions because they can compute a value for each row in the input group. They are particularly useful for computing moving averages, a cumulative sum, or the rank of each row.

- The Catalyst optimizer enables you to write efficient data processing applications. The cost-based optimizer was introduced in Spark 2.2 to enable Catalyst to be more intelligent about selecting the right kind of join implementation based on the collected statistics of the processed data.

- Project Tungsten is the workhorse behind the scenes that speeds up the execution of data process applications by employing a few advanced techniques to improve the efficiency of using memory and CPU.

CHAPTER 5

Optimizing Spark Applications

Chapter 4 covered major capabilities in Spark SQL to perform simple to complex data processing. When you use Spark to process large datasets in hundreds of gigabytes or terabytes, you encounter interesting and challenging performance issues; therefore, it is important to know how to deal with them. Mastering Spark application performance issues is a very interesting, challenging, and broad topic. It requires a lot of research and a deep understanding of some of the key areas of Spark related to memory management and data movement.

A comprehensive tuning guide is a very large surface area and deserves to be in its own book. However, this is not the intent of this chapter. Rather, it aims to discuss a few common Spark application performance issues that you might encounter in your journey of developing Spark applications.

First, it describes a set of common performance issues. Then it goes into detail about common techniques for improving Spark application performance, such as tuning important Spark configurations and leveraging in-memory computation. The last part looks at the intelligent optimization techniques in the Adaptive Query Execution (AQE) framework introduced in Spark 3.0, such as dynamically coalescing shuffle partitions, dynamically switching join strategies, and optimizing skew joins. These techniques enable Spark developers to focus more time on building powerful Spark applications and less time optimizing the performance.

Common Performance Issues

If you do a quick Internet search on common Spark application performance issues, you discover they broadly fall into two categories: OOM (out of memory) and taking a long time to complete. To overcome these performance issues, it requires understanding the

© Hien Luu 2021
H. Luu, *Beginning Apache Spark 3*, https://doi.org/10.1007/978-1-4842-7383-8_5

underlying mechanisms about how Spark handles memory management and how it performs some of the complex and expensive transformations such as joining, grouping, and aggregations.

Spark Configurations

One of the important aspects of optimizing Spark application performance is to know which knobs are available to use, how to apply those knobs, and when those knobs are effective. In the Spark world, these knobs are known as Spark properties. There are hundreds of them, and most have reasonable default values. For a comprehensive list of Spark properties, you can review the Spark property documentation at `https://spark.apache.org/docs/latest/configuration.html`.

There are two important aspects of Spark properties that every Spark application developer needs to know: the three different ways of setting properties and the two different kinds of properties.

Different Ways of Setting Properties

There are three different ways of setting properties, which are described in the order of precedence.

The first way is done through a set of configuration files. Under the directory where you installed Spark, there is a `conf` folder with three files: `log4j.properties.template`, `spark-env.sh.template`, and `spark-default.conf.template`. Simply adding the desired properties and associated values and then saving them without the `.template` suffix let Spark know to apply those properties to the Spark cluster and all the Spark applications submitted to the cluster. In other words, the properties in these configuration files are applicable at the global cluster level.

The second way is to specify the Spark properties when submitting your Spark application via the `spark-submit` or start up a Spark shell via the `spark-shell` command line using `--config` or `-c` flag. Listing 5-1 is an example of passing in Spark properties via the command line.

Listing 5-1. Passing in Spark Properties via Command Line

```
./bin/spark-submit --conf "spark-executor-memory=4g" --class org.apache.
spark.examples.SparkPi ./examples/jars/spark-examples_2.12-3.1.1.jar
```

The third way is to directly specify Spark properties through the `SparkConf` object in your Spark application. Listing 5-2 is a very short example of setting properties in a Spark application in Scala.

Listing 5-2. Setting Spark Properties Directly in a Spark Application in Scala

```scala
import org.apache.spark.sql.SparkSession
def main(args: Array[String]) {
  val sparkSession = SparkSession.builder
                      .config("spark.executor.memory","4g")
                      .config("spark.eventLog.enabled","true")
                      .appName("MyApp").getOrCreate()
}
```

Since there are multiple ways of setting properties, Spark establishes the following precedence. The properties set directly on the `SparkConf` object take highest precedence, followed by the flags passed on the `spark-submit` or `spark-shell` command line, and then the options in the configuration file.

Different Kinds of Properties

Not all Spark properties are created equally. Understanding which ones are meant to be used at which part of your application deployment life cycle is important. Spark properties can mainly be divided into two kinds: deployment and runtime.

The deployment-related properties are set once during the Spark application launching step, and they are not meant to be changed after that. Therefore, it is fruitless to set them programmatically through the `SparkConf` object at runtime. Examples of this type of property are `spark.driver.memory` or `spark.executor.instances.` It is recommended to set these properties either through the configuration files or through the `spark-submit` command line.

The runtime-related properties control the various aspects of Spark during the running of your Spark application, such as `spark.sql.shuffle.partitions`, the number of partitions to use when shuffling data for joins and aggregations. In addition to setting these in configuration files through the command line, they can also be set programmatically and repeatedly through the `SparkConf` object.

Viewing Spark Properties

After setting the Spark properties, it is important to verify their values. One of the easiest ways to view Spark properties is in the Environment tab of the Spark web UI. Figure 5-1 shows an example of the Spark properties.

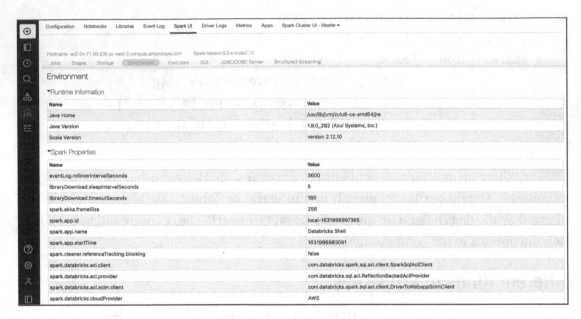

Figure 5-1. *Spark properties in the Environment tab*

Another easy way to view Spark properties is by programmatically retrieving them from the SparkConf object in your Spark application. Listing 5-3 shows how to do that.

Listing 5-3. Displaying Spark Properties Programmatically

```scala
scala> for (prop <- spark.conf.getAll.keySet) {
         println(s"${prop}: ${spark.conf.get(prop)}")
     }

spark.sql.warehouse.dir: file:/<path>/spark-warehouse
spark.driver.host: 192.168.0.22
spark.driver.port: 63834
spark.repl.class.uri: spark://<ip>:63834/classes
```

```
spark.repl.class.outputDir:/private/var/folders/_m/nq53ddp...
spark.app.name: Spark shell
spark.submit.pyFiles:
spark.ui.showConsoleProgress: true
spark.app.startTime: 1621812866358
spark.executor.id: driver
spark.submit.deployMode: client
spark.master: local[*]
spark.home: /<path>/spark-3.1.1-bin-hadoop2.7
spark.sql.catalogImplementation: hive
spark.app.id: local-1621812867559
```

Spark Memory Management

One of the challenges Spark developers run into when developing and operating Spark applications that process large amounts of data is dealing with out-of-memory (OOM) errors. When this happens, your Spark application stops working or crashes and the only thing you can do is figure out the underlying cause, fix it and restart your application.

Before going into the OOM issue, let's step back to examine how Spark manages its memory on the driver and the executor. As mentioned in Chapter 1, each Spark application consists of one driver and one or more executors.

Spark Driver

Regarding memory management, the amount of memory a Spark driver is determined by the spark.driver.memory configuration, which is usually specified when starting a Spark application via the spark-submit command or a similar mechanism in the Databricks cluster creation process. The default value for the spark.driver.memory configuration is 1 GB.

The driver is mainly responsible for orchestrating Spark applications' workload, so it doesn't need as much memory as executors. There are two scenarios when it needs to allocate memory, and any incorrect usage of Spark in these scenarios likely causes an OOM issue.

The first scenario is when either the RDD.collect or DataFrame.collect action is invoked. This action transfers all the data of the RDD or DataFrame from all the executors over to the application's driver, which then tries to allocate the necessary memory to store the transferred data. For example, if the data size in the DataFrame is

187

about 5 GB and the driver has only 2 GB, you encounter the OOM issue when calling actions. The closest analogy to this situation is dumping water from five one-gallon buckets into a single two-gallon bucket.

The Spark documentation recommends these actions should only be used when the data amount is expected to be small and can fit into the driver memory. Two common ways to reduce the size of an RDD or DataFrame is to perform some sort of filtering and calling the limit transformation before collecting the data to the Spark driver side.

The second scenario is when Spark is trying to broadcast the content of a dataset. It does this either because your Spark application uses broadcasting variables by calling the `SparkContext.broadcast` function or while Spark is performing a broadcast join, which is discussed in the "Understanding Spark Joins" section.

The broadcasting variable is an optimizing technique where Spark sends an immutable copy of the dataset from the driver to all the executors in a Spark application. Once an executor has a copy of the dataset, it is available to all the current and future tasks on that executor to consume. Essentially, this is to avoid transferring the same dataset multiple times to executors when it is needed repeatedly. To ensure you don't run into the OOM issue, make sure the size of your dataset is small or increase the memory size of the driver.

While performing a broadcast join, Spark sends the data of the smaller DataFrame of the two being joined to all the executors. Like the broadcasting variable situation, you encounter an OOM issue if the data of the smaller DataFrame is too big. You can control when Spark use the broadcast join by setting the value of the `spark.sql.autoBroadcastJoinThreshold` configuration, which has a default value of 10 GB. If you want to disable the broadcast join, you can set this configuration value to –1.

Spark Executor

The memory management on the Spark executor side is more involved and complicated than on the driver side. First, let's discuss how it manages memory and what it decides to store in the memory.

At runtime, each executor is a JVM process running on an operating system. The amount of memory available is allocated in a Java virtual machine (JVM) memory heap. When launching a Spark application, you can specify the JVM memory heap size for an executor by specifying a value for the `spark.executor.memory` property, which has a default value of 1 GB. The JVM heap is divided into three areas: Spark memory, user memory, and reserved memory (see Figure 5-2).

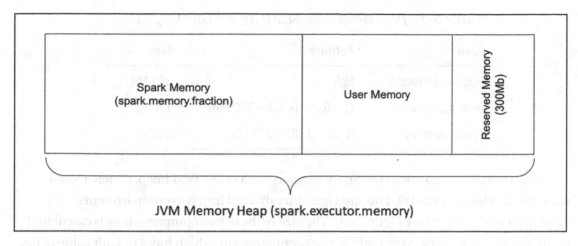

Figure 5-2. *The different areas of JVM heap*

Figure 5-2 indicates not all the memory specified in the spark.executor.memory property is allocated to the executor. It is important to keep that in mind when dealing with memory-related issues in your Spark application. Let's explore each area in the JVM memory heap.

- **Reserved memory** sets aside a fixed amount of memory for its internal usage. As of Spark version 3.0, the reserved amount is 300 MB.

- **User memory** is the area for storing objects created from a Spark application developer's data structure, the internal metadata in Spark, and safeguarding against OOM errors in case of sparse and usually large records. The size of this area is calculated as (1 – spark. memory.fraction) * (spark.executor.memory – reserved memory).

- **Spark memory** is the area under the control of the executor. It is used for execution and storage. The size of this area is calculated as (spark.memory.fraction) * (spark.executor.memory – reserved memory).

To make it more concrete, let's walk through an example where the JVM heap size (spark.executor.memory) is 4 GB and uses the default value of the spark.memory. fraction configuration as 0.6. Table 5-1 lists the size of each of the JVM heap areas.

Table 5-1. *JVM Heap Area Size with a 4 GB Heap Size*

Area	Formula	Size
Reserved memory	N/A	300 MB
User memory	(1 - 0.6) * (4 GB – 300 MB)	1.4 GB
Spark memory	(0.6) * (4 GB – 300 MB)	2.2 GB

Now let's dive deeper into the Spark memory area of the JVM heap, which a Spark executor has total control of. This area is further divided into two compartments: execution and storage (see Figure 5-3). The size of these two compartments is calculated using the `spark.memory.storageFraction` configuration, which has a default value of 0.5.

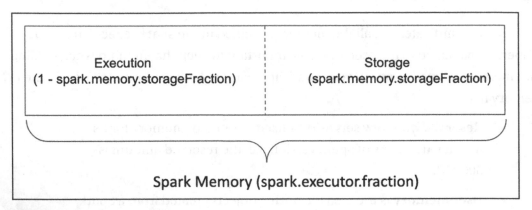

Figure 5-3. *The two compartments in the Spark memory area*

The Spark executor decides what to store in each compartment and it can expand and contract each compartment as needed. The following lists what is stored in each compartment and how its initial size is calculated.

- Execution

  - This compartment is used for buffering intermediate data during the execution of a task such as shuffling data, joining two datasets, performing aggregations, or sorting data. These objects are often short-lived and no longer needed after the task is completed.

  - The size of this compartment is calculated as (Spark memory area * (1 – `spark.memory.storageFraction`))

- Storage

  - This compartment is used for storing all the cached data for future and repeated accesses and broadcast variables. When you invoke either the persist() or cache() API on a DataFrame, its data is persisted in memory and is called cached data. The Spark application user controls the lifetime of the cached data, but the data could be evicted due to memory pressure.

  - The size of this compartment is controlled by spark.memory. storageFraction and is calculated as (Spark memory area * spark.memory.storageFraction)

The Unified Memory Management feature was introduced in Spark version 1.6 to intelligently manage these two compartments in a way that works for most workloads, requires very little expertise in tuning the memory fractions, and takes advantage of the unused storage memory where applications don't cache data that much.

It achieves these goals by instituting a boundary between the compartments as a flexible barrier where it can be moved to either side based on the need and establishing specific rules to avoid memory starvation and system failure. The rules are as follows.

- When execution memory exceeds its compartment, it can borrow as much of the storage memory as is free.

- When storage memory exceeds its compartment, it can borrow as much of the execution memory as is free.

- When execution needs more memory and some of its memory was borrowed by the storage compartment, it can forcefully evict that memory occupied by storage.

- When storage needs more memory and some of its memory is borrowed by the execution compartment, it cannot forcefully evict that memory occupied by execution. Storage must wait until the executor releases them.

The main motivations behind these rules are to provide execution as much space as it needs, and evicting storage memory is significantly simpler to implement.

To prevent the storage memory starvation, there needs to be a guarantee about the minimum reserved memory for cached data. This is accomplished by requiring the executor to honor the amount of memory specified by the `spark.memory.storageFraction` property, which specifies the amount of storage memory that is immune from eviction. In other words, the storage memory for cached data may only be evicted if total storage exceeds the storage compartment size.

The rules and guarantee work well for Spark to support multiple types of workloads. When a particular type of workload performs a lot of complex and wide transformations, and it doesn't need to cache much data, it can leverage the free memory in the storage compartment as much as it needs.

Table 5-2 describes the memory-related properties.

Table 5-2. *Memory-Related Spark Properties*

Property Name	Default Value	Description
`spark.driver.memory`	1GB	Amount of memory to use for the driver process
`spark.executor.memory`	1 GB	Amount of memory to use per executor process
`spark.memory.fraction`	0.6	Fraction of (JVM heap space – 300 MB) used for execution and storage. The lower. It is recommended to leave this at the default value.
`spark.memory.storageFraction`	0.5	The amount of storage memory is immune to eviction, expressed as a fraction of the region's size set aside by `spark.memory.fraction`. The higher it is, the less memory might be available for execution and tasks may spill to disk more often. It is recommended to leave this at the default value

Next, let's discuss a few common scenarios where you might run into the OOM issue.

Leverage In-Memory Computation

One of the distinguishing features of Spark from other data processing engines or frameworks is the ability to perform in-memory computation to speed up the data processing logic in a Spark application. This section discusses when to leverage this unique feature and how to persist data in Spark temporary storage (either memory or disk) across the executors in your Spark cluster.

When to Persist and Cache Data

Spark provides the ability to persist (or cache) the data in your DataFrame in memory, which is available to any future operations to use. As a result, those operations run much faster, often by 10x, because the data they need is read from the computer memory. This capability is very useful for iterative algorithms and when the data of a DataFrame is needed or reused multiple times. Machine learning algorithms are highly iterative, meaning they run through many iterations to produce an optimized model. This capability is also useful to provide fast and interactive use of the data for analytical purposes. In short, when the data of a DataFrame is reused multiple times in your Spark applications, it is highly recommended that you consider persisting that DataFrame data to speed up your Spark application.

Let's walk through a simple example to illustrate the need for leveraging in-memory computation. Imagine you are asked to analyze a large log file with billions of rows to identify the root cause of a recent production issue by analyzing the various types of exceptions. After reading the data from the log file into a DataFrame, the next step is to filter those rows down to only the rows that contain the word *exception*. Then, persist those rows in memory so you can repeatedly analyze the different kinds of exceptions.

Persisting a dataset does incur some costs in terms of serialization, deserialization, and storage cost. If a dataset is only to be used once, caching slows down your Spark application.

Persistence and Caching APIs

Spark DataFrame class provides two APIs to persist data: `cache()` and `persist()`. Both offer the same capability. The former is a shorthand version of the latter, which provides more control over how and where your DataFrame should be stored, such as in memory or on disk, and whether the data is stored in a serialized format.

It is reasonable to ask what happens when your Spark application doesn't have sufficient memory to cache the large dataset in memory. For example, let's say your Spark application has ten executors, and each one has 6 GB of RAM. If the size of a DataFrame you want to persist in memory is 70 GB, it wouldn't fit into 60 GB of RAM. This is where the storage-level concept comes in. There are two options you can choose from when persisting the data: location and serialization. The location determines whether the data should be stored in memory, on disk, or a combination of the two. The serialization option determines whether data should be stored as a serialized object or not. These two options represent the different types of trade-offs you are making: CPU time and memory usage. Table 5-3 describes the different options, and Table 5-4 describes the trade-off information.

Table 5-3. *Storage and Serialization for Persisting Data*

Storage Level	Description
MEMORY_ONLY	Persist data as deserialized objects in memory only
MEMORY_AND_DISK	Persist data as deserialized objects in memory. If there isn't sufficient memory, the rest is stored as serialized objects on a disk.
MEMORY_ONLY_SER	Persist data as serialized objects in memory only.
MEMORY_AND_DISK_SER	Similar to MEMORY_AND_DISK, but persist data as serialized objects in memory.
DISK_ONLY	Persist data as serialized objects on disk only.
MEMORY_ONLY_2, MEMORY_AND_DISK_2	The same as MEMORY_ONLY and MEMORY_AND_DISK, but replicate the data on two cluster nodes.

Table 5-4. *Memory Space vs. CPU Time Trade-offs*

Storage Level	Memory Space	CPU Time
MEMORY_ONLY	High	Low
MEMORY_AND_DISK	High	Medium
MEMORY_ONLY_SER	Low	High
MEMORY_AND_DISK_SER	Low	High
DISK_ONLY	Low	High

When the data of a DataFrame is no longer needed to be persisted, you can use the unpersist() API to remove it from memory or disk depending on the specified level when calling the persist() API. A spark cluster has limited memory; if you keep persisting DataFrames, Spark uses the LRU eviction policy to automatically evict the persisted ones that have not been recently accessed when the available memory amount is low.

Persistence and Caching Example

This section walks through an example of persisting the data in Spark show performance improvement. It also looks at an example Spark UI to show the partitions stored in a Spark executor. The example in Listing 5-4 first generates the app_log_df DataFrame with 300 million rows with three columns: id, message, date. Then it filters app_log_df DataFrame to contain only messages with the word *exception* in the message column and finally assigns the output to the except_log_df DataFrame.

Listing 5-4. Code Example to Persisting Data in Spark

```
import org.apache.spark.sql.functions._
import scala.util.Random

val log_messages = Seq[String](
    "This is a normal line",
    "RuntimeException - this is really bad",
    "ArrayIndexOutOfBoundsException - don't do this",
    "NullPointerException - this is a nasty one",
    "SQLException - bad SQL again!!!"
)

// set up functions and UDF
def getLogMsg(idx:Int) : String = {
    val randomIdx = if (Random.nextFloat() < 0.3) 0 else
    Random.nextInt(log_messages.size)
    log_messages(randomIdx)
}
val getLogMsgUDF = udf(getLogMsg(_:Int):String)
```

```
// generate a DataFrame
val app_log_df = spark.range(30000000).toDF("id")
                  .withColumn("msg", getLogMsgUDF(lit(1)))
                  .withColumn("date",
                              date_add(current_timestamp,
                              -(rand() * 360).cast("int")))

// generate a DataFrame with "Exception" message
val except_log_df = app_log_df
                           .filter(msg.contains("Exception"))

// before persisting
except_log_df.count()

// call persist transformation to persist exception_log_df
except_log_df.cache()

// materialize the exception_log_df
except_log_df.count()

// this should be really fast
// since exception_log_df is now in memory
exception_log_df.count()

// evict exception_log_df from memory
exception_log_df.unpersist()
```

Figure 5-4 shows the duration for each of the three count actions. Job 0 is for the first call, which took 8 seconds. Job 0 is for the second call, which forces Spark to materialize DataFrame data in memory, which took 7 seconds. Job 2 is for the last call, which took only 56 milliseconds. As you can see, accessing the data in memory is extremely fast.

▾ Completed Jobs (3)

Page: 1 — 1 Pages. Jump to 1 . Show 100 items in a page. Go

Job Id ▾	Description	Submitted	Duration	Stages: Succeeded/Total	Tasks (for all stages): Succeeded/Total
2	count at <console>:30 count at <console>:30	2021/09/18 13:42:44	65 ms	2/2	13/13
1	count at <console>:30 count at <console>:30	2021/09/18 13:42:32	7 s	2/2	13/13
0	count at <console>:30 count at <console>:30	2021/09/18 13:42:13	7 s	2/2	13/13

Page: 1 — 1 Pages. Jump to 1 . Show 100 items in a page. Go

Figure 5-4. *Shows the duration of each of the count action*

To see how Spark stores the except_log_df DataFrame in memory, bring up the Spark UI and navigate to the Storage tab. Figure 5-5 shows the number of cached partitions and the total size in memory.

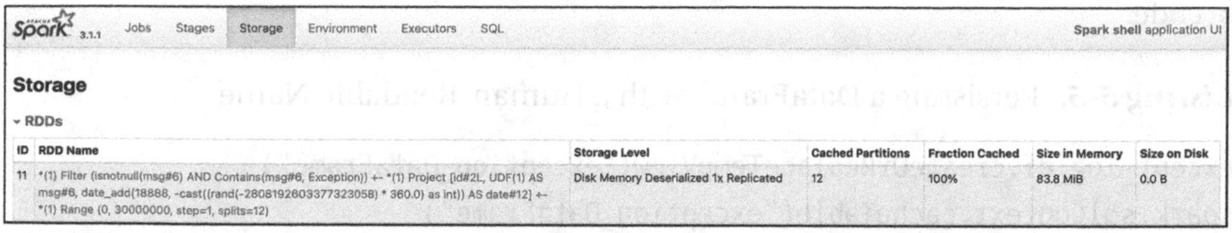

Figure 5-5. *The Storage tab in Spark UI shows the cached DataFrame*

To see the detailed information of each partition of the persisted DataFrame, click the link under the RDD name column. Figure 5-6 shows each partition's size, location, storage level, and replication factor.

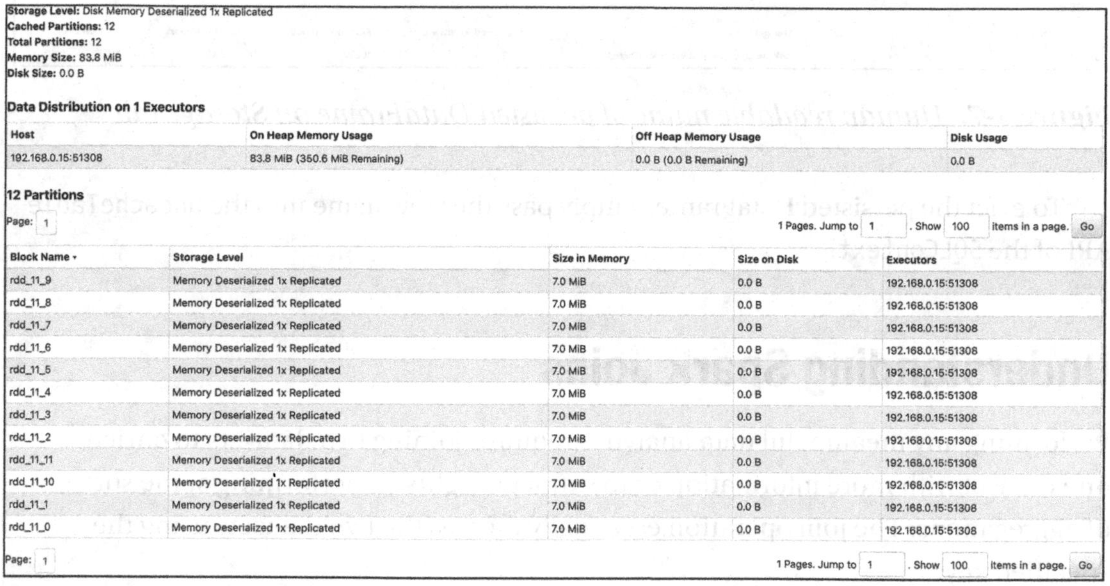

Figure 5-6. *The detailed information of the cached partitions*

You can specify the storage and replication factor by calling the persist API with an appropriate argument of type StorageLevel.

To programmatically evict the persisted DataFrame, you can simply call the unpersist API. The partitions disappear from memory and the Storage tab of Spark UI.

Note that the RDD Name column in Figure 5-5 has a long and cryptic name. To specify a human-readable name of a persisted DataFrame, you must take a few extra lines of code. The first step is to create a temporary view with a name and then pass that same name into the `cachTable` API of `SQLContext` use. Listing 5-5 shows the needed line of code.

Listing 5-5. Persisting a DataFrame with a Human-Readable Name

```
except_log_df.createOrReplaceTempView("exception_DataFrame")
spark.sqlContext.cacheTable("exception_DataFrame")
```

If those two lines were executed successfully then the name under the RDD Name column of the Storage tab should look something like Figure 5-7.

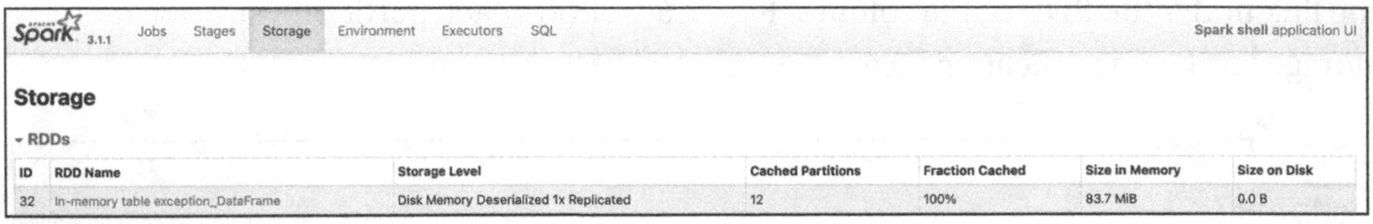

ID	RDD Name	Storage Level	Cached Partitions	Fraction Cached	Size in Memory	Size on Disk
32	In-memory table exception_DataFrame	Disk Memory Deserialized 1x Replicated	12	100%	83.7 MiB	0.0 B

Figure 5-7. *Human readable name of persisted DataFrame on Storage tab*

To evict the persisted DataFrame, simply pass the view name into the `uncacheTable` API of the `SQLContext`.

Understanding Spark Joins

Performing any meaningful data analytics requires joining two datasets to enrich one dataset with more information or to extract insights by performing some sort of aggregations. The join operation essentially merges the two datasets using the specified key.

Spark has extensive support for join operation and the various join types in DataFrame, Dataset, and Spark SQL APIs. This is covered in Chapter 4. This chapter discusses some of the join strategies that Spark uses and memory and performance-related aspects.

A join strategy is an approach to carry out a join operation. Spark supports five different join strategies: broadcast hash join (BHJ), shuffle hash join (SHJ), shuffle sort-merge join (SMJ), broadcast nested loop join (BNLJ), and shuffle-and-replicated nested loop join (a.k.a. cartesian product join). A summary of each is captured in Table 5-5. Out of these five strategies, the commonly used ones in Spark are BHJ and SMJ because they are versatile enough to handle most of the scenarios, such as when the size of one of the datasets is small or when the size of both datasets are large.

Table 5-5. *Join Strategy Summary*

Name	Join Condition	Description
broadcast hash join	equi joins	When one of the datasets is small enough to broadcast across the network, perform the hash join.
shuffle hash join	equi joins	When two datasets are large, shuffle them across the network, then perform the hash join.
shuffle merge join	equi joins	When two datasets are large, shuffle them across the network, sort them and then merge.
broadcast nested loop join	equi and non-equi joins	Broadcast the smaller dataset and use the nested loops to perform the join.
shuffle-replicated nested loop join	equi and non-equi joins	Cartesian product join. Very slow.

Broadcast Hash Join

Among the five join strategies, BHJ, also known as the map-side-only join, is the simplest and fastest in Spark and is applicable when one of the datasets is small. This strategy broadcasts a copy of the smaller dataset via the driver to all executors in the Spark cluster. Then the join is performed against the larger dataset, as shown in Figure 5-7.

Spark uses the value of the property `spark.sql.autoBroadcastJoinThreshold` to determine which dataset is eligible for broadcasting and its default value is 10 MB. If your Spark cluster has sufficient memory to handle the broadcasting of a larger dataset, you can simply increase the threshold to an appropriate value. BHJ is depicted in Figure 5-8.

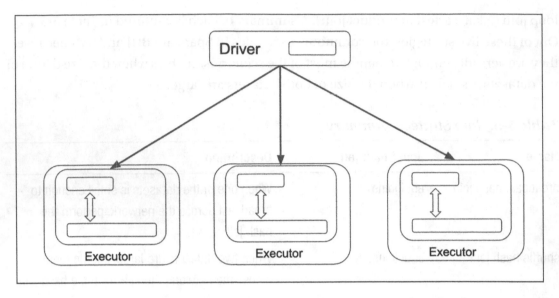

Figure 5-8. *Broadcasting smaller dataset from driver to executors hash join strategy*

When you perform a join between two datasets and are certain the size of one of them is smaller than the threshold of the property mentioned, you can verify that Spark uses the BHJ strategy by examining the execution plan. Listing 5-6 shows a simple example of joining two datasets.

Listing 5-6. Joining One Small Dataset with a Larger One

```
import org.apache.spark.sql.functions._
val small_df = Seq(("WA", "Washington"), ("CA", "California"),
                   ("AZ", "Arizona"), ("AK", "ALASKA"))
              .toDF("code", "name")
val large_df = spark.range(500000).toDF("id")
          .withColumn("code", when(rand() < 0.2, "WA")
                             .when(rand() < 0.4, "CA")
                             .when(rand() < 0.6, "AZ")
                             .otherwise("AK"))
          .withColumn("date",
             date_add(current_date,
                     -(rand() * 360).cast("int")))

val joined_df = small_df.join(large_df, "code")
```

Spark UI has a good visual way to example the execution plan, as depicted in Figure 5-9. Simply navigate to the SQL tab and click the job you want to execute. This example shows the smaller dataset is broadcasted via the BroadcastExchange operator, and then the BroadHashJoin strategy is employed.

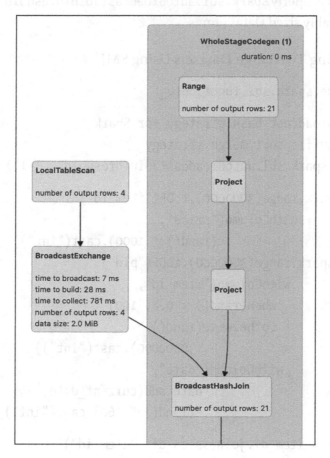

Figure 5-9. *Broadcasting hash join strategy in Spark UI*

Shuffle Sort Merge Join

Spark's other commonly used join strategy is SMJ, an efficient approach to joining two large datasets. SMJ first shuffles the rows with the same key from both datasets to the same partition on the same executor. Next, those rows are sorted by the specified join key. Finally, the merge step iterates over those rows and merges the ones with matching keys. The sort-merge process is similar to the one in the merge sort algorithm. It is an efficient way of merging two large datasets without loading one into memory first, like the one in SHJ.

Listing 5-7 shows a small example of simulating SMJ by first disabling the SHJ by setting the value of property `spark.sql.autoBroadcastJoinThreshold` to –1, and then joining two reasonably sized DataFrames.

Listing 5-7. Joining Two Large Datasets Using SMJ

```
import org.apache.spark.sql.functions._

// disable the broadcast hash strategy for Spark
// to use the shuffle sort merge strategy
spark.conf.set("spark.sql.autoBroadcastJoinThreshold", "-1")

val item_df = spark.range(3000000).toDF("item_id")
                    .withColumn("price",
                                (rand() * 1000).cast("int"))
val sales_df = spark.range(3000000).toDF("pId")
                    .withColumn("item_id",
                        when(rand() < 0.8, 100)
                        .otherwise(rand() *
                                    30000000).cast("int"))
                    .withColumn("date",
                                date_add(current_date,
                                -(rand() * 360).cast("int")))

val item_sale_df = item_df.join(sales_df, "item_id")

item_sale_df.show()

+-------+-----+-------+----------+
|item_id|price|    pId|      date|
+-------+-----+-------+----------+
|  18295|  484|2607123|2020-09-27|
|  19979|  261|1121863|2020-07-05|
|  37282|  915|1680173|2020-10-04|
|  54349|  785| 452954|2021-05-28|
|  75190|  756| 142474|2021-02-19|
```

```
|  89806|  763|1842105|2020-06-26|
|  92357|  689|1451331|2021-03-07|
| 110753|  418|1803550|2020-11-25|
| 122965|  729| 917035|2020-06-22|
| 150285|  823|2306377|2020-10-05|
| 180163|  591| 330650|2020-07-13|
| 181800|  606|2065247|2020-11-06|
| 184659|  443| 982178|2020-09-01|
| 198667|  796|2985859|2021-04-02|
| 201833|  464| 709169|2020-07-31|
| 208354|  357| 927660|2021-05-30|
| 217616|  627| 174367|2021-04-25|
| 223396|  752|2850510|2020-11-05|
| 225653|  188|2439243|2021-01-16|
| 233633|  628|2811113|2020-12-02|
+-------+-----+-------+----------+
```

After running the code, you can navigate to the SQL tab in Spark UI to see the execution plan, which looks like Figure 5-10. Notice the sorting step takes place first, and then next is the merging step.

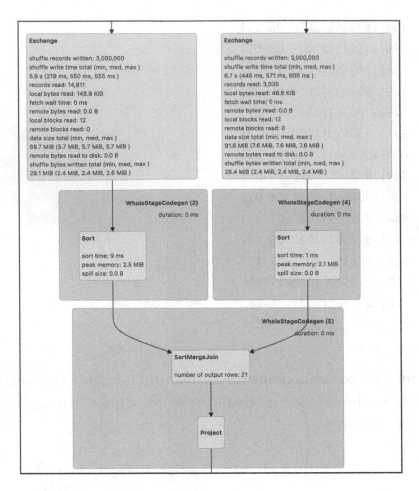

Figure 5-10. *Shuffle merge join strategy in Spark UI*

Adaptive Query Execution

Among the useful and innovative features introduced in Spark 3.0 is the Adaptive
Query Execution (AQE) framework. It should be a favorite feature of Spark application
developers who develop and maintain large data pipelines that process hundreds
of terabytes to petabytes or data analysts who use Spark daily to perform complex
and interactive analysis on large datasets. AQE extends Spark SQL's query optimizer
and planner to dynamically adjust and regenerate high-quality and optimized query
execution plans using the latest statistics about row count, partition size, and such to
automatically address most of the common performance issues and to speed up Spark
application completion time or prevent them from running into OOM.

In Spark 2.0, the Spark SQL Catalyst optimizer provides a generic framework for generating query execution plans with a set of rule-based optimizations to improve performance. Cost-based optimization was introduced in Spark 2.2 to improve the optimizations by leveraging the generated per column data statistics such as cardinality, min/max values, and NULL values.

AQE was introduced in Spark 3.0 to leverage the runtime statistics about the partitions between the stages to adjust and regenerate the query execution plans to improve performance of some of the most encountered performance scenarios, such as too many partitions, skew data, or using the wrong kind of joins.

To better understand how AQE works, let's first revisit a few core concepts about Spark jobs, stages, and tasks. When an action type API of a DataFrame is called, such as count(), or collect(), or show(), Spark launches a job to carry out that action by executing all the previous transformations that led up to it.

Each job consists of one or more stages, and a stage is needed whenever Spark encounters a wide transformation, which involves moving its input data from multiple partitions. Examples of wide transformations include groupBy() or orderBy(). Each stage materializes its intermediate result to disk, and the following stage can only proceed if the materialization of all the partitions is completed. This represents a natural place for AQE to adapt its execution plan by leveraging the partitions' runtime statistics, such as the total number of partitions and the size of each one. That is the main reason why the first word in the name of each AQE feature starts with the word *dynamically*. One thing to note, AQE is helpful only when your Spark applications have one or more stages.

Figure 5-11 visually summarizes the flow described in this paragraph.

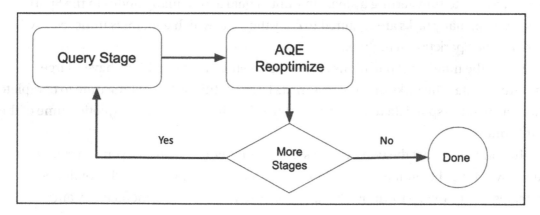

Figure 5-11. *AQE in action*

The performance optimizations provided in AQE are extremely easy to use. Simply just turn on the appropriate properties, and AQE leaps into action. The top-level property to enable AQE is called `spark.sql.adaptive.enabled`, so make sure to set it to true before trying out the following examples. Each feature has its own set properties to control the various fine-grained aspects of that particular feature, which are discussed in each section.

AQE framework provides these three extremely useful features.

- Dynamically coalescing shuffle partitions

- Dynamically switching join strategies

- Dynamically optimizing skew joins

The following sections discuss what they are and how they work.

Dynamically Coalescing Shuffle Partitions

One of the most expensive operations in Spark is shuffling, meaning moving the data across the network such that the data is redistributed appropriately based on the needs of subsequent operations. If not tuned accordingly, this operation can significantly impact query performance. The main tuning knob about shuffling is the number of partitions. The query performance degrades if the number is either too low or too high. A common challenge is the best number of partitions is often specific to the use case at hand. It might need to be updated frequently due to the data volume increase or any changes to the data processing logic.

When the number of partitions is too large, it creates unnecessary inefficiency about I/O usage due to transferring a small amount of data across many nodes in the Spark cluster. Also, many tasks are required to copy the data, which generates unnecessary work for the Spark task scheduler.

When the number of partitions is too small, some of the partitions have a large amount of data. The tasks that process those large partitions take a long time to complete and may need to spill data to disk. As a result, this slows down the completion time of the whole query.

To combat against the two extremes of the number of partitions, you can start your query by setting the number of partitions to be large. AQE automatically combines or coalesce the small partitions into larger ones to overcome the inefficiency and performance-related issues described by leveraging the most up-to-date statistics at the end of a shuffle. Figure 5-12 visually depicts an example where AQE is not enabled, and those small partitions are processed separated reducers. Figure 5-13 visually depicts an example where AQE is enabled. The small partitions (B, C, D) coalesce into a single partition, and therefore the overall number of reducers is less. The figures are inspired by this Databricks log (see https://databricks.com/blog/2020/05/29/adaptive-query-execution-speeding-up-spark-sql-at-runtime.html).

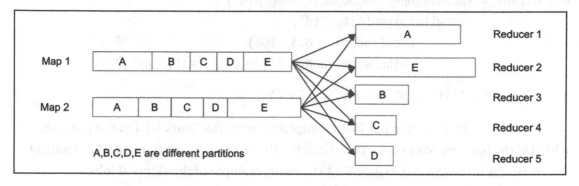

Figure 5-12. *AQE is disabled*

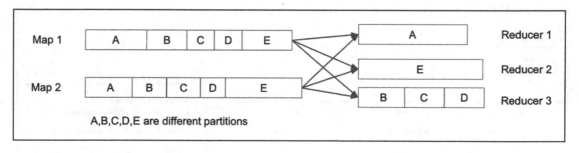

Figure 5-13. *AQE is enabled*

Now let's see the dynamically coalescing shuffle partitions feature in action by simulating a scenario to analyze the count of a small transactional fact table with approximately 80% of the items sold is a popular item with an ID of 100. Listing 5-8 generates 100 million rows, and 80% of the item_ids with a value of 100. Next, it counts each item_id and then counts the different item_ids.

Listing 5-8. Performing the Analysis of Item Count from a Synthetic Small Transaction Fact Table

```
import org.apache.spark.sql.functions._
// make sure the AQE is enabled
spark.conf.set("spark.sql.adaptive.enabled", "true")
spark.conf.get("spark.sql.adaptive.enabled")

val dataDF = spark.range(100000000).toDF("pId")
                .withColumn("item_id",
                    when(rand() < 0.8, 100)
                    .otherwise(rand() * 3000000).cast("int"))

dataDF.groupBy("item_id").count().count()
```

Once the count() action has been completed, go to the Spark UI. On the Jobs tab, which is the first one, you notice two out of the three jobs were skipped, and the bulk of the work was done in job 0. Figure 5-14 shows an example of the skipped jobs.

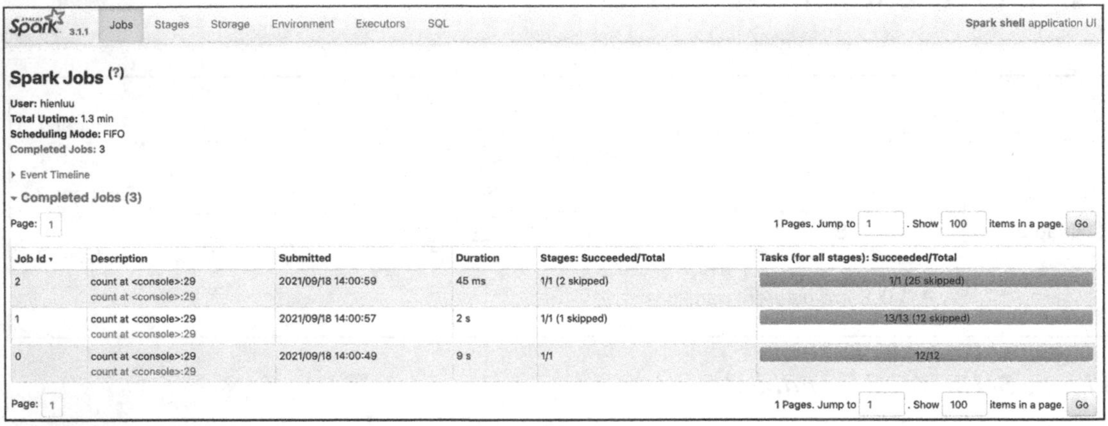

Figure 5-14. *Spark UI shows skipped jobs*

Let's examine job 2 by clicking the hyperlink under the Description column. The Spark UI shows stages 2 and 3 were skipped, and you see an operator called CustomShuffleReader operator at the top of stage 2. The AQE framework introduced this operator to coalesce partitions with the size that is smaller than the size specified by the property spark.sql.adaptive.advisoryPartitionSizeInBytes, which has a default value of 64 MB. Figure 5-15 shows the Spark UI with skipped stages.

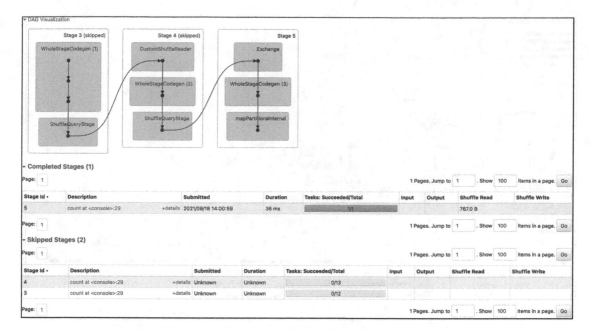

Figure 5-15. *Spark UI shows skipped jobs the skipped stages*

If the AQE is disabled, you would see a very different execution plan which consists of a single job (see Figure 5-16) with three stages (see Figure 5-17). The most notable part is the second stage partition the data into 200 partitions, according to the default value of the property `spark.sql.shuffle.partitions`, where the majority of them have only a small amount of data, as depicted in Figure 5-18.

Figure 5-16. *Spark UI shows a single job to perform GroupBy operator*

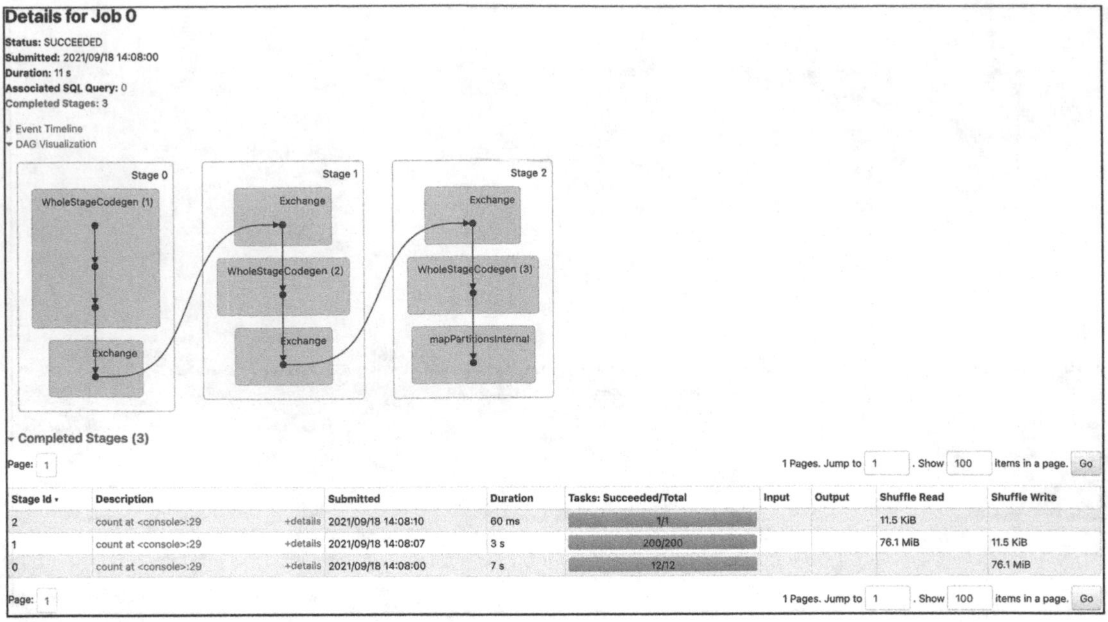
Details for Job 0
Status: SUCCEEDED
Submitted: 2021/09/18 14:08:00
Duration: 11 s
Associated SQL Query: 0
Completed Stages: 3
Event Timeline
DAG Visualization
Stage 0
WholeStageCodegen (1)
Exchange
Stage 1
Exchange
WholeStageCodegen (2)
Exchange
Stage 2
Exchange
WholeStageCodegen (3)
mapPartitionsInternal
Completed Stages (3)
Page: 1
1 Pages. Jump to 1 . Show 100 items in a page. Go
Stage Id | Description | Submitted | Duration | Tasks: Succeeded/Total | Input | Output | Shuffle Read | Shuffle Write
2 | count at <console>:29 +details | 2021/09/18 14:08:10 | 60 ms | 1/1 | | | 11.5 KiB |
1 | count at <console>:29 +details | 2021/09/18 14:08:07 | 3 s | 200/200 | | | 76.1 MiB | 11.5 KiB
0 | count at <console>:29 +details | 2021/09/18 14:08:00 | 7 s | 12/12 | | | | 76.1 MiB
Page: 1
1 Pages. Jump to 1 . Show 100 items in a page. Go

Figure 5-17. *Spark UI shows three stages with their tasks*

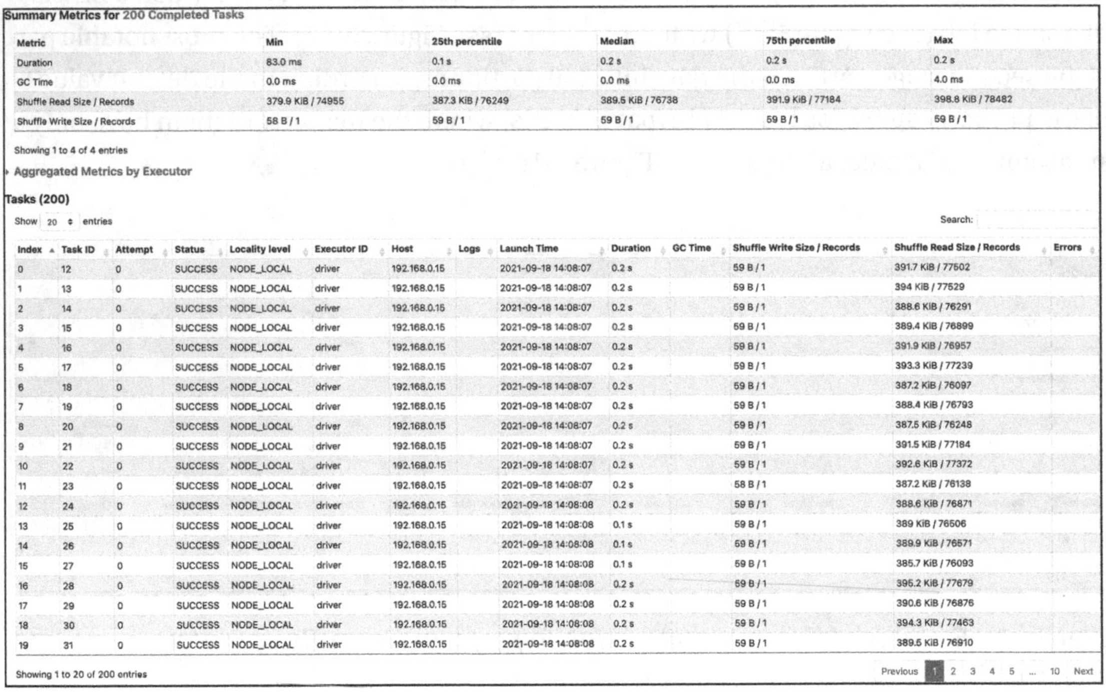
Summary Metrics for 200 Completed Tasks
Metric | Min | 25th percentile | Median | 75th percentile | Max
Duration | 83.0 ms | 0.1 s | 0.2 s | 0.2 s | 0.2 s
GC Time | 0.0 ms | 0.0 ms | 0.0 ms | 0.0 ms | 4.0 ms
Shuffle Read Size / Records | 379.9 KiB / 74955 | 387.3 KiB / 76249 | 389.5 KiB / 76738 | 391.9 KiB / 77184 | 398.8 KiB / 78482
Shuffle Write Size / Records | 58 B / 1 | 59 B / 1 | 59 B / 1 | 59 B / 1 | 59 B / 1
Showing 1 to 4 of 4 entries
Aggregated Metrics by Executor
Tasks (200)
Show 20 entries
Search:
Index | Task ID | Attempt | Status | Locality level | Executor ID | Host | Logs | Launch Time | Duration | GC Time | Shuffle Write Size / Records | Shuffle Read Size / Records | Errors
0 | 12 | 0 | SUCCESS | NODE_LOCAL | driver | 192.168.0.15 | | 2021-09-18 14:08:07 | 0.2 s | | 59 B / 1 | 391.7 KiB / 77502 |
1 | 13 | 0 | SUCCESS | NODE_LOCAL | driver | 192.168.0.15 | | 2021-09-18 14:08:07 | 0.2 s | | 59 B / 1 | 394 KiB / 77529 |
2 | 14 | 0 | SUCCESS | NODE_LOCAL | driver | 192.168.0.15 | | 2021-09-18 14:08:07 | 0.2 s | | 59 B / 1 | 388.6 KiB / 76291 |
3 | 15 | 0 | SUCCESS | NODE_LOCAL | driver | 192.168.0.15 | | 2021-09-18 14:08:07 | 0.2 s | | 59 B / 1 | 389.4 KiB / 76899 |
4 | 16 | 0 | SUCCESS | NODE_LOCAL | driver | 192.168.0.15 | | 2021-09-18 14:08:07 | 0.2 s | | 59 B / 1 | 391.9 KiB / 76957 |
5 | 17 | 0 | SUCCESS | NODE_LOCAL | driver | 192.168.0.15 | | 2021-09-18 14:08:07 | 0.2 s | | 59 B / 1 | 393.3 KiB / 77239 |
6 | 18 | 0 | SUCCESS | NODE_LOCAL | driver | 192.168.0.15 | | 2021-09-18 14:08:07 | 0.2 s | | 59 B / 1 | 387.2 KiB / 76097 |
7 | 19 | 0 | SUCCESS | NODE_LOCAL | driver | 192.168.0.15 | | 2021-09-18 14:08:07 | 0.2 s | | 59 B / 1 | 388.4 KiB / 76793 |
8 | 20 | 0 | SUCCESS | NODE_LOCAL | driver | 192.168.0.15 | | 2021-09-18 14:08:07 | 0.2 s | | 59 B / 1 | 387.5 KiB / 76248 |
9 | 21 | 0 | SUCCESS | NODE_LOCAL | driver | 192.168.0.15 | | 2021-09-18 14:08:07 | 0.2 s | | 59 B / 1 | 391.5 KiB / 77184 |
10 | 22 | 0 | SUCCESS | NODE_LOCAL | driver | 192.168.0.15 | | 2021-09-18 14:08:07 | 0.2 s | | 59 B / 1 | 392.6 KiB / 77372 |
11 | 23 | 0 | SUCCESS | NODE_LOCAL | driver | 192.168.0.15 | | 2021-09-18 14:08:07 | 0.2 s | | 58 B / 1 | 387.2 KiB / 76138 |
12 | 24 | 0 | SUCCESS | NODE_LOCAL | driver | 192.168.0.15 | | 2021-09-18 14:08:08 | 0.2 s | | 59 B / 1 | 388.6 KiB / 76871 |
13 | 25 | 0 | SUCCESS | NODE_LOCAL | driver | 192.168.0.15 | | 2021-09-18 14:08:08 | 0.1 s | | 59 B / 1 | 389 KiB / 76506 |
14 | 26 | 0 | SUCCESS | NODE_LOCAL | driver | 192.168.0.15 | | 2021-09-18 14:08:08 | 0.1 s | | 59 B / 1 | 389.9 KiB / 76571 |
15 | 27 | 0 | SUCCESS | NODE_LOCAL | driver | 192.168.0.15 | | 2021-09-18 14:08:08 | 0.1 s | | 59 B / 1 | 385.7 KiB / 76093 |
16 | 28 | 0 | SUCCESS | NODE_LOCAL | driver | 192.168.0.15 | | 2021-09-18 14:08:08 | 0.2 s | | 59 B / 1 | 395.2 KiB / 77679 |
17 | 29 | 0 | SUCCESS | NODE_LOCAL | driver | 192.168.0.15 | | 2021-09-18 14:08:08 | 0.2 s | | 59 B / 1 | 390.6 KiB / 76876 |
18 | 30 | 0 | SUCCESS | NODE_LOCAL | driver | 192.168.0.15 | | 2021-09-18 14:08:08 | 0.2 s | | 59 B / 1 | 394.3 KiB / 77463 |
19 | 31 | 0 | SUCCESS | NODE_LOCAL | driver | 192.168.0.15 | | 2021-09-18 14:08:08 | 0.2 s | | 59 B / 1 | 389.5 KiB / 76910 |
Showing 1 to 20 of 200 entries
Previous 1 2 3 4 5 ... 10 Next

Figure 5-18. *Spark UI shows 200 tasks*

There is a small set of properties for the dynamically coalesced shuffle partition feature to control its behavior. Table 5-6 describes these properties. Their names start with the same `spark.sql.adaptive` prefix, so it is omitted for brevity.

Table 5-6. *Dynamically Coalesce Shuffle Partitions Properties*

Property	Default Value	Description
<prefix>.coalescePartitions. enabled	True	Fine-grain control to enable partition coalescing.
<prefix>advisoryPartition SizeInBytes	64 MB	The approximate target coalesced partition size. The size is not larger than the target size.
<prefix>.coalescePartitions. minPartitionSize	1 MB	The minimum size of coalesced partitions. The size is not smaller than this value.
<prefix>.coalescePartitions. minPartitionNum	default Spark parallelism when not specified	The suggested minimum number of shuffle partitions after coalescing.

Dynamically Switching Join Strategies

Among the different joint types Spark supports, the most performant one is called broadcast hash join. In the past, Spark developers would either need to give hints to Spark or explicitly request Spark to use broadcast hash join by marking one of the datasets using the `broadcast` function. With AQE enabled, when Spark detects the size of one side of the joint below the broadcast-size threshold, it dynamically switches to broadcast hash join to speed up the query performance. This is useful when one of the joint datasets starts with a large size, and after the filter operator is done, its size is dramatically reduced. AQE has the updated statistics of all the partitions between stages, and therefore it can leverage it to determine whether it makes sense to switch join at runtime.

The threshold to control whether the broadcast hash join should be used is the `spark.sql.autoBroadcastJoinThreshold` property, which has a default value of 10 MB. If the data size of one of the datasets in a join is less than that value, then AQE switch to broadcast hash join. It makes sense to increase the threshold to a higher value if your Spark cluster has sufficient memory to store the broadcasted dataset in memory. If you are using Databricks, then the property is called `spark.databricks.adaptive.autoBroadcastJoinThreshold`.

Now let's see the dynamically switching join strategies feature in action by simulating a scenario where after joining two datasets, the aggregation logic has a filtering condition such that the size of one dataset fall below the autobroadcast joint threshold. AQE can detect this at runtime, and it switches the sort-merge join to the broadcast hash join to speed up the query. The initial static plan is not aware of the selectivity of the filter.

Listing 5-9 sets up two DataFrames—car registration and car sales. A filter is applied to one of the columns after they are joined. The filter condition is not known at the initial execution plan.

Listing 5-9. Apply a Filter Condition after Joining Car Registrations and Car Sales DataFrames

```
spark.conf.set("spark.sql.adaptive.enabled", true)
import org.apache.spark.sql.functions._
import scala.util.Random

// setting up functions and UDFs
def getCarByIdx(idx:Int) : String = {
  val validIdx = idx % popularCars.size
  val finalIdx = if (validIdx == 0) 1 else validIdx
  popularCars(finalIdx)
}

def randomCar(idx:Int): String = {
  val randomCarIdx = if (Random.nextFloat() < 0.4) 0 else
                       Random.nextInt(popularCars.size)
  popularCars(randomCarIdx)
}

val getCarByIdxUDF = udf(getCarByIdx(_:Int):String)
val randomCarUDF = udf(randomCar(_:Int):String)

// setting the data frames
val car_registration_df = spark.range(5000000).toDF("id")
        .withColumn("make", when('id > 1,
```

```
        getCarByIdxUDF('id)).otherwise("FORD:F-Series"))
    .withColumn("sale_price",
            (rand() * 100000).cast("int"))

val car_sales_df = spark.range(30000000).toDF("id")
        .withColumn("make",randomCarUDF(lit(5)))
        .withColumn("date",
            date_add(current_date,
                    -(rand() * 360).cast("int")))

car_sales_df.join(car_registration_df, "make")
        .filter('sale_price < 300)
        .groupBy("date").agg(
                sum("sale_price").as("total_sales"),
                count("make").as("count_make"))
        .orderBy('total_sales.desc)
        .select('date, 'total_sales).show()
```

Dynamically Optimizing Skew Joins

Skew joins are one of the most annoying performance issues when joining two datasets in parallel data computation systems like Spark. The data skew in one of the joined datasets causes the imbalance in the partition data size. As a result, the join takes a long time to complete. The data skew in real-life datasets is a natural phenomenon and occurs more often than you think due to popularity, population concentration, or consumer behavior. For example, the population of West Coast and East Coast cities tend to be larger than other US cities. In past years, Spark application developers have come up with many innovative ideas to overcome the skew join performance issue, requiring additional effort. This feature only speeds up the skew joins. It requires very little effort from Spark developers. Undoubtedly, this feature put a big smile on a lot of Spark developers.

Note   From the probability theory and statistics field of studies, skewness is a measure of the symmetry of a distribution. In the data processing world, the data skew suggests the distribution is uneven or asymmetric, where some column values have way more rows than others, and some columns have very few.

To perform a join, Spark assigns a partition id to each row of joined datasets based on the hash of the joined keys and then shuffle those rows with the same partition id to the same partition. If one of the datasets has data skew, then one or more partitions have way more rows than others, as depicted in Figure 5-19. The task assigned to process the large partition takes much longer to complete than other tasks that process smaller partitions.

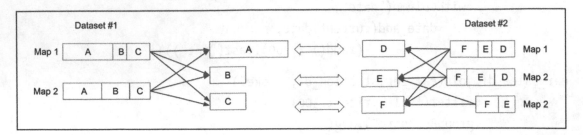

Figure 5-19. *Skew partitions*

When AQE is enabled, the large partition is split into a few smaller ones to be processed by multiple tasks in parallel to speed up the overall join completion time. Using the example in Figure 5-9, partition A, the skewed one, is split into two smaller partitions. Partition D is duplicated two times, as depicted in Figure 5-20. As a result, four tasks are running the join, and each one roughly takes about the same amount of time; therefore, the overall completion time is shorter.

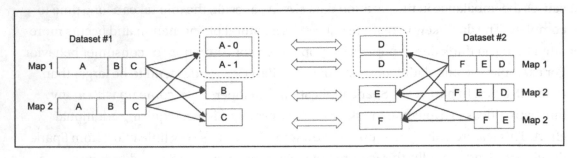

Figure 5-20. *Skew partition is split, and corresponding partitions are duplicated*

Now let's see the dynamically optimizing skew join feature in action by simulating a scenario where one of the two joined datasets has skew data and see how much AQE speeds up the join by. Listing 5-10 is an adaptation of an example from the blog at `https://coxautomotivedatasolutions.github.io/datadriven/spark/data%20skew/ joins/data_skew/`.

Listing 5-10 demonstrates the skew join optimization by setting up two datasets, car registrations, and car sales. There is a data skew in the second one; 40% of the sales are for Ford F-Series by calling the `randomCarUDF` function.

Listing 5-10. Joining Car Registrations and Car Sales DataFrames (Cars Sales Has Skew Data)

```
import org.apache.spark.sql.functions._

// make sure to turn on AQE
spark.conf.set("spark.sql.adaptive.enabled", true)

// since the data size is small, avoid broadcast hash join,
// reduce the skew partition factor and threshold
spark.conf.set("spark.sql.autoBroadcastJoinThreshold", -1)
spark.conf.set("spark.sql.adaptive.skewJoin.
              skewedPartitionFactor", "1")
spark.conf.set("spark.sql.adaptive.skewJoin.
              skewedPartitionThresholdInBytes", "2mb")

val popularCars = Seq[String]("FORD:F-Series",
        "RAM:1500/2500/3500","CHEVROLET:SILVERADO",
        "TOYOTA:RAV4","HONDA:CRV",
        "TOYOTA:TACOMA","HONDA:CIVIC",
        "TOYOTA:COROLLA","GMC:SIERRA"
)

// setting up functions and UDFs
def getCarByIdx(idx:Int) : String = {
  val validIdx = idx % popularCars.size
  val finalIdx  = if (validIdx == 0) 1 else validIdx
  popularCars(finalIdx)
}

def randomCar(idx:Int): String = {
  val randomCarIdx = if (Random.nextFloat() < 0.4) 0 else
  Random.nextInt(popularCars.size)
  popularCars(randomCarIdx)
}
```

```scala
val getCarByIdxUDF = udf(getCarByIdx(_:Int):String)
val randomCarUDF = udf(randomCar(_:Int):String)

// create the two data frames to join
val car_registration_df = spark.range(500000).toDF("id")
          .withColumn("registration",
              lit(Random.alphanumeric.take(7).mkString("")))
          .withColumn("make",
                    when('id > getCarByIdxUDF('id))
                    .otherwise("FORD:F-Series"))
          .withColumn("engine_size",
                      (rand() * 10).cast("int"))

val car_sales_df = spark.range(30000000).toDF("id")
            .withColumn("make",randomCarUDF(lit(5)))
            .withColumn("engine_size",
                        (rand() * 11).cast("int"))
            .withColumn("sale_price",(
                        rand() * 100000).cast("int"))
            .withColumn("date",
                        date_add(current_date,
                        - rand() * 360).cast("int")))

// perform the join by make
car_registration_df.join(car_sales_df, "make")
                .groupBy("date").agg(
                    sum("sale_price").as("total_sales"),
                    count("make").as("count_make"))
                .orderBy('total_sales.desc)
                .select('date, 'total_sales).show()
```

From the partition statistics, AQE sees the skew data in the car sales DataFrame and split the large partition into three smaller partitions. Figure 5-21 shows the details of the split in the execution plan.

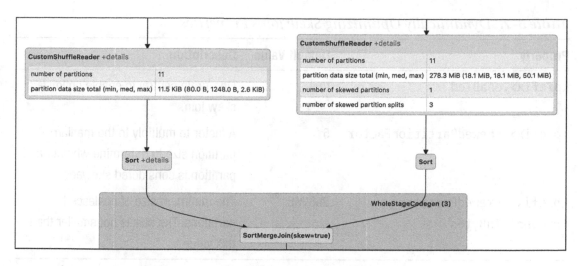

Figure 5-21. *One skew partition split into three partitions*

Once the skewed partition is split into smaller ones, both the median and max duration are 41 seconds. Figure 5-22 shows the task completion time.

Metric	Min	25th percentile	Median	75th percentile	Max
Duration	5 s	7 s	41 s	41 s	43 s
GC Time	2 s	2 s	17 s	19 s	19 s
Spill (memory)	0.0 B	0.0 B	160.5 MiB	160.5 MiB	160.5 MiB
Spill (disk)	0.0 B	0.0 B	35.2 MiB	35.2 MiB	35.2 MiB
Shuffle Read Size / Records	12.2 MiB / 959584	28.2 MiB / 2219184	37.3 MiB / 4520059	37.3 MiB / 4521387	37.3 MiB / 4522239
Shuffle Write Size / Records	13.1 KiB / 360	13.1 KiB / 360	13.2 KiB / 360	13.2 KiB / 360	13.2 KiB / 360

Summary Metrics for 14 Completed Tasks

Figure 5-22. *Task duration even out at 41 seconds*

This feature also has a small set of properties to control its behavior. Table 5-7 describes the properties. Their names start with the same `spark.sql.adaptive.skewJoin` prefix, so it is omitted for brevity.

Table 5-7. *Dynamically Optimizing Skew Joins Properties*

Property	Default Value	Description
`<prefix>.enabled`	true	Fine-grain control to enable optimize a skew join.
`<prefix>.skewedPartitionFactor`	5	A factor to multiply to the median partition size to determine whether a partition is considered skewed.
`<prefix>.skewedPartition ThresholdInBytes`	256 MB	The minimum size of coalesced partitions. The size is not smaller than this value.

A partition is considered skewed when both of the following conditions are true.

- partition size > `skewedPartitionFactor` * median partition size
- partition size > `skewedPartitionThresholdInBytes`

Summary

Spark application tuning and optimization is a broad topic. This chapter covered common performance-related challenges, including memory issues and long-running query performance. The ability to optimize Spark applications requires a broad understanding of some of the tuning knobs, how Spark manages its memory, and taking advantage of some of the new capabilities in AQE.

- The tuning knobs in Spark are the properties, and there are three different ways of setting them. The first way is through the configuration file. The second way is to specify them on the command line when launching your Spark applications. The final way is to set them programmatically in your Spark applications, which takes the highest precedence. Spark UI provides an easy way to view the configured properties, organized under the Environment tab.

- Memory is another important resource. A Spark driver uses its allocated memory to store broadcast variables and the data collected from all Spark executors, which flexibly manages its allocated memory to handle different workloads.

- One of the unique features of Spark is the ability to perform in-memory computation, which can speed iterative and interactive use cases by 10x. If a dataset is used multiple times in a Spark application, it is a good candidate to persist its data in memory or on disk.

- Introduced in Spark 3.0, the Adaptive Query Execution framework can perform three different optimizations at runtime by leveraging the up-to-date statistics about the materialized partitions between stages to speed up your query performance. The first one is about coalescing shuffle partitions to improve I/O efficiency and reduce the burden on the Spark scheduler. The second one is about dynamically switching a join to broadcast hash join from a sort-merge join when the size of one of the datasets falls below the broadcast-size threshold. The last one is about optimizing skew joins by detecting and splitting skewed partitions into multiple smaller partitions to speed up the over query performance.

CHAPTER 6

Spark Streaming

In addition to batch data processing, stream processing has become a must-have capability for any business to harness the value of real-time data to increase their competitive advantages, make better business decisions, or improve user experience. With the advent of the Internet of Things, the volume and velocity of real-time data has increased. For Internet companies like Facebook, LinkedIn, or Twitter, millions of social activities happening every second on their platform are represented as streaming data.

At a high level, stream processing is about the continuous processing of unbounded streams of data. Doing this at scale in a fault-tolerant and consistent manner is a challenging task. Luckily, stream processing engines such as Spark, Flink, Samza, Heron, and Kafka have been steadily and dramatically matured over the last few years to enable businesses to build and operate complex stream processing applications much easier than before.

More interesting real-time data processing use cases have emerged as the community understands how best to apply the increasingly matured streaming engines to their business needs. For example, Uber leverages stream processing capabilities to understand the number of riders and drivers on their platform in real time. These near real-time insights influence business decisions like moving excess drivers from a low-demand area to higher-demand areas in a city.

Most Internet companies leverage an experimentation system to perform A/B testing when releasing new features or trying new designs. Stream processing enables a faster reaction to the experiments by reducing the time it takes to understand the experiment effectiveness from days to hours.

Fraud detection is an area that has embraced stream processing with open arms due to the benefits it gains from instant insights of fraud activities, so they can either be stopped or monitored. For large companies that have hundreds of online services, a common need is to monitor their health by processing the large volume of generated logs in near real-time via stream processing. There are many more interesting real-time data processing use cases, and some of them are shared in this chapter.

© Hien Luu 2021

H. Luu, *Beginning Apache Spark 3*, https://doi.org/10.1007/978-1-4842-7383-8_6

This chapter starts with describing useful stream processing concepts and then provides a short introduction to the stream processing engine landscape. Then the remaining sections of this chapter describe the Spark stream processing engine in detail and its APIs.

Stream Processing

In the world of big data, batch data processing is widely known since the introduction of Hadoop. The popular MapReduce framework is one of the components in the Hadoop ecosystem, and it became the king of batch data processing because of its capabilities and robustness. After a period of innovation in the batch data processing area, most of the challenges in this space are now well understood. Since then, the big data open source community has shifted its focus and innovations to the stream processing space.

Batch data processing applies the computational logic through a fixed size and static input dataset and produces the result at the end. This means the processing stop when it gets to the end of the dataset. By contrast, stream processing is about running the computational logic through unbounded streams of data, and therefore the processing is continuous and long-running. Although the difference between batch data and streaming data is mainly about the finiteness, stream processing is much more complex and challenging than batch data processing because of the unbounded data nature, the incoming order of real-time data, the different rates that the data arrive, and the expectation of the correctness and low latency in the face of machine failure.

In the world of batch data processing, it is common to hear that it takes hours to finish a complex batching data processing job because of the size of the input datasets.

There is an expectation that stream processing engines must provide low latency and high throughput by delivering incoming streams of data as efficiently as possible to applications, so they can react to or extract insight quickly. Performing any interesting and meaningful stream processing usually involves maintaining a state in a fault-tolerant manner. For example, a stock trading streaming application wants to maintain and display the top 10 or 20 most actively traded stocks throughout the day. To accomplish this goal, the running count of each stock must be maintained by the stream processing engine on behalf of the application or by the application itself. Usually, the state is maintained in memory and backed by resilient storage like disk, which is resilient to machine failures.

Stream processing doesn't work in a silo. Sometimes there is a need to work together with batch data processing to enrich the incoming streaming data. A good example of this is when a page view streaming application needs to compute page view statistics of its users based on user location; then, it needs to join user clicks data with member data. A good stream processing engine should provide an easy way to join batch data with streaming data without much effort.

One of the common use cases of stream processing is to perform some aggregations of incoming data and then write that summarized data out to an external data sink to be consumed by either a web application or a data analytics engine. The desire here is to have an end-to-end exactly-once guarantee of the data in the face of failure, whether because of machine failures or some bugs in the data processing application. The key here is how does the stream processing engine deal with failure such that the incoming data is not lost and not double-counted.

As stream processing engines mature, they provide fast, scalable, and fault-tolerant distributed system properties and developer-friendly ways to perform data streaming computation by up-leveling an abstraction from low-level APIs to high-level declarative language as SQL. With this advancement, it is much easier to build a self-service streaming platform to enable product teams to quickly make meaningful business decisions by tapping into the data or events generated by various company products. Remember, one of the goals in data stream processing is to extract business insights in a timely manner so businesses can either react quickly or take business actions.

In summary, stream processing has its own set of unique challenges resulting from processing data that is continuous and unbounded. It is important to be mindful of these challenges as you set out to build long-running stream processing applications or evaluate a particular stream processing engine. The challenges are as follows.

- Reliably maintaining potentially large state for data streaming applications

- Efficiently and quickly deliver messages for applications to process

- Dealing with streaming data that arrives out of order

- Joining with batch data to enrich the incoming streaming data

- End-to-end exactly once guarantee delivery of data even where there is failure

- Dealing with uneven data arrival rate

Concepts

To perform stream processing, it is imperative to understand the following core and useful concepts. These important concepts are very much applicable to developing streaming applications on any stream processing engine. Knowing them is useful to evaluating stream processing engines; they also enable you to ask the right questions to find out how much support a particular stream processing engine provides in each of these areas.

- Data delivery semantics
- Notion of time
- Windowing

Data Delivery Semantics

When a piece of data enters a stream processing engine, it is responsible for delivering it to the streaming application for processing. There are three types of guarantees that a stream processing engine can provide even under failure scenarios.

- **At most once**: This implies a stream processing engine guarantees that a piece of data is delivered to an application no more than one time, but it could be zero time. In other words, there is a chance that a piece of data is lost, and therefore the application does not see it at all. For some use cases, this is acceptable, but it is not for some other use cases. One of those use cases is the financial transaction processing applications. Losing data can result in not charging customers and, therefore, a reduction in revenue.

- **At least once**: This implies a stream processing engine guarantees that a piece of data is delivered to an application one or more times. There is no data lost in this case; however, there is a potential for double or triple counting. In the example of the financial transaction processing applications, a transaction is applied multiple times, resulting in customer complaints. This guarantee is stronger than at most once because no data is lost.

- **Exactly once**: This implies a stream processing engine guarantees that a piece of data is delivered to an application exactly one time only, no more and no less. In this case, there is no data loss and no double counting. Most modern and popular stream processing engines provide this kind of guarantee. Of the three guarantees, this one is the most desirable one for building critical business streaming applications.

One way of looking at these delivery semantics is that they fall into a spectrum, where at most once is the weakest guarantee and exactly once is the strongest guarantee, depicted in Figure 6-1.

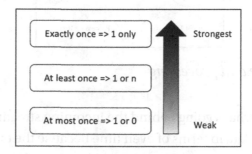

Figure 6-1. *Delivery semantics spectrum*

When evaluating a stream processing engine, it is important to understand the level of guarantee it provides and the implementation behind this guarantee. Most modern stream processing engines employ a combination of check-pointing and write-ahead logs techniques to provide an exactly-once guarantee.

Notion of Time

In the world of stream processing, the notion of time is very important because it enables you to understand what's going on in terms of time. For example, in the case of a real-time anomaly detection application, the notion of time gives insights into the number of suspicious transactions occurring in the last 5 minutes or a certain part of the day.

There are two important types of time: event time and processing time. As depicted in Figure 6-2, event time represents when the piece data was created, and this information is typically encoded in the data. For example, in IoT devices that take ocean temperature in a certain part of the world, the event time is when the temperature was taken. The encoding of the temperature data may consist of the temperature itself and timestamp. The processing time represents the time when the stream processing engine processes a piece of data. In the example of the ocean temperature IoT devices, the processing time is the clock time of the stream processing engine when it starts to perform transformations or aggregations on the temperature data.

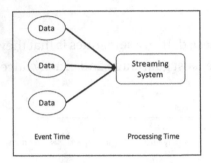

Figure 6-2. *Event time and processing*

To truly understand what's going behind the incoming stream of data, it is imperative to process the incoming data in terms of even time because the event time represents the point in time that the data was created. In an ideal state, the data arrive and be processed shortly after it was created, and therefore the gap between the event time and the processing time is short. That is often not the case, and therefore the lag varies over time according to the conditions that prevent the data from arriving immediately after they are created. The greater the lag, the greater the need to process data using event time and not using the processing time.

Figure 6-3 illustrates the relationship between event time and processing time, and an example of what a real lag looks like. The notion of time is very much related to the windowing concept, which is described next. To deal with unbounded incoming streams of data, one common practice in the stream processing engines is to divide the incoming data into chunks by using the start and end time as the boundary. It makes more sense to use event time as the temporal boundaries.

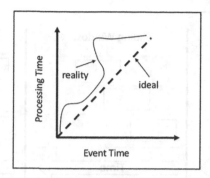

Figure 6-3. *The lag between event time and processing time*

Windowing

Given the unbounded nature of streaming data, it is not feasible to have a global view of the incoming streaming data. Hence, to extract any meaningful value from the incoming data, you must process them in chunks. For example, given a traffic count sensor that emits a count of the number of cars every 20 seconds, it is not feasible to compute a final sum. Instead, it is more logical to ask how many cars pass that sensor every minute or five minutes. In this case, you need to partition the traffic counting data into chunks of 1 minute or 5 minutes, respectively. Each chunk is called a window.

Windowing is a common stream processing pattern where the unbounded coming streaming of data is divided into chunks based on temporal boundaries—either event time or processing time. Although the former is used more commonly to reflect the actual reality of the data. However, given that the data may not arrive in the order they were created or delayed due to network congestion, it is impossible to always have all the data created in that time window.

There are three commonly used windowing patterns, and most modern stream processing engines support them. The three patterns are depicted in Figure 6-4.

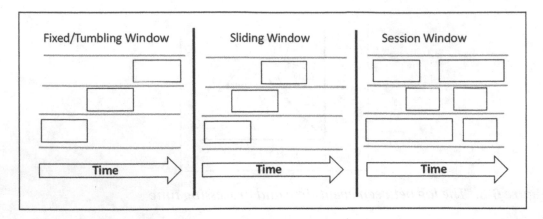

Figure 6-4. *Three commonly used windowing patterns*

Fixed/tumbling window divides the incoming stream of data into fixed-size segments, where each one has a window length, a start time, and an end time. Each coming piece of data is slotted into one and only one fixed/tumbling window. With this small batch of data in each window, it is easy to reason when performing aggregations like sum, max, or average.

A sliding window is another way of dividing the incoming stream of data into fixed-size segments, where each one has a window length and sliding interval. If the sliding interval is the same size as the window length, it is the same as the fixed/tumbling window. The example in Figure 6-4 shows the sliding interval is smaller than the window length. This implies that one or more pieces of data are included in more than one sliding window. Because of the overlapping of the windows, the aggregation produces a smoother result than in the fixed/tumbling window.

The session window type is commonly used to analyze user behavior on a website. Unlike the fixed/tumbling and sliding window, it has no predetermined window length. Rather, it is usually determined by a gap of inactivity that is greater than some threshold. For example, the length of a session window on Facebook is determined by the duration of activities that a user does, such as browsing the user feeds, sending messages, and so on.

Stream Processing Engine Landscape

There is no shortage of innovations from the open source community in coming up with solutions for stream processing. In fact, there are multiple options to choose a stream processing engine from. Some of the earlier stream processing engines were born out of necessity, some later ones were born out of research projects, and some evolved from batching processing engines. This section presents a few popular stream processing engines: Apache Storm, Apache Samza, Apache Flink, Apache Kafka Streams, Apache Apex, and Apache Beam.

Apache Storm is one of the pioneers in stream processing, and its popularity is mainly associated with the large-scale stream processing that Twitter does. Apache Storm's initial release was in 2011, and it became the Apache top-level project in 2014. In 2016, Twitter abandoned Apache Storm and switched over to Heron, which is the next generation of Apache Storm. Heron is more resource-efficient and provides much better throughput than Apache Storm.

Apache Samza was born at LinkedIn to help solve its stream processing needs, and it was open sourced in 2013. It was designed to work very closely with Kafka and runs on top of Hadoop YARN for process isolation, security, and fault tolerance. Apache Samza was designed to process streams, which are composed of ordered, partitioned, replayable, and fault-tolerant sets of immutable messages.

Apache Flink started as a fork of the research project called "Stratosphere: Information Management on the Cloud." It became an Apache top-level project in 2015, and ever since then, it has been steadily gaining popularity as a high-throughput and low latency streaming engine. One key difference between Apache Flink and Apache Storm and Apache Samza is that it supports both batch and stream processing in the same engine.

Apache Kafka has evolved from a distributed publish-subscribe messaging system to a distributed streaming platform. It was created at LinkedIn and became a top-level Apache project in 2012. Unlike other stream processing engines, Kafka's stream processing capabilities are packaged as a lightweight library, making it very easy to write real-time streaming applications.

Apache Apex is a relative newcomer to this space. It was developed by a company called DataTorrent, and they decided to open source it in 2016. Apache Apex is considered a Hadoop YARN native platform that unifies stream and batch processing.

Apache Beam is quite an interesting project that came out of Google in 2016. The main idea behind this project is to provide a common layer of powerful and easy-to-use abstraction for both streaming and batch processing that is portable across a variety of runtime platforms (i.e., Apache Flink, Apache Spark, Google Cloud DataFlow). In other words, think of Apache Beam as an uber-API for big data processing.

There are two standard stream processing models, and each of the stream processing engines (except Apache Beam) is subscribed to one of them. The two models are called *record-at-a-time* and *micro-batching*, which are depicted in Figure 6-5.

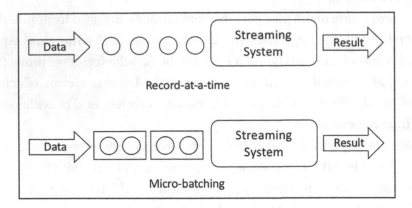

Figure 6-5. *Two different models of stream processing*

Both models have inherent advantages and disadvantages. The record-at-a-time model does what it sounds like. It immediately processes each piece of incoming piece of data as it arrives. As a result, this model can provide low latency in milliseconds. The micro-batching model waits and accumulates a small batch of input data based on a configurable batching interval and processes each batch in parallel. The micro-batching model can't provide the same level of latency as the other model. In terms of throughput, the micro-batch has a much higher rate because a batch of data is processed in an optimized manner, and therefore the cost per piece of data is low compared to the other model. One interesting side note is that it is fairly easy to build a micro-batching model on top of the record-of-a-time model.

Of all the stream processing engines discussed, only Apache Spark employs the micro-batching model; however, some work is already underway to support the record-at-a-time model.

Spark Streaming Overview

One of the contributing factors to the popularity of Apache Spark's unified data processing platform is the ability to perform stream processing and batch data processing.

With the high-level description of the intricacies and challenges of stream processing and a few core concepts out of the way, the remainder of this chapter focuses on the Spark streaming topic. First, it provides a short and high-level understanding and some of the capabilities of Spark's first-generation stream processing engine called DStream. Then the bulk of the remaining chapter provides information about Spark's second stream processing engine called Structured Streaming.

New Spark streaming applications should be developed on top of Structured Streaming to take advantage of some of the unique and advanced features it provides.

Spark DStream

The first generation of Spark stream processing engine was introduced in 2012, and the main programming abstraction in this engine is called *discretized stream,* or DStream. It works by employing the micro-batching model to divide the incoming stream of data into batches, which are then processed by the Spark batch processing engine. This makes a lot of sense at the time when RDD was the main programming abstraction model. Each batch is internally represented by an RDD. The number of pieces of data in a batch is a function of the incoming data rate and the batch interval. Figure 6-6 visually describes the way DStream works at a high level.

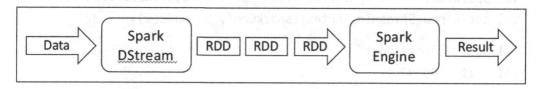

Figure 6-6. *Spark DStream*

A DStream can be created from an input data stream from Kafka, AWS Kinesis, a file, or a socket. One of the key pieces of information needed when creating a DStream is the batch interval, which can be in seconds or milliseconds. With a DStream, you can apply high-level data processing functions such as map, filter, reduce, or reduceByKey on the incoming stream of data. Additionally, you can perform windowing operations such as reducing and counting over either a fixed/tumbling or a sliding window by providing a window length and a sliding interval. One important note is that the window length and sliding interval must be multiples of the batch interval. For example, if the batch interval is three seconds and the fixed/tumbling interval is used, the window length and sliding interval can be six seconds. Although, maintaining an arbitrary state while performing computations across batches of data is supported in DStream, it is a manual process and a bit cumbersome. One of the cool things you can do with a DStream is to join it with another DStream or an RDD representing static data. After all the processing logic is complete, you can use DStream to write out the data to external systems such as a database, a file system, or HDFS.

New Spark streaming applications should be developed on the second-generation Spark stream processing engine, called Structured Streaming, which is covered in the next section. For the remainder of this section, you look at a small word count Spark DStream application; the goal is to understand what a typical Spark DStream application looks like. Listing 6-1 contains the code for the word count application, which is an example from Apache Spark source code GitHub repository (see `https://bit. ly/2G8N30G`).

Listing 6-1. Apache Spark DStream Word Count application

```
object NetworkWordCount {
  def main(args: Array[String]) {

    // Create the context with a 1 second batch size
    val sparkConf = new SparkConf().setAppName("NetworkWordCount")
    val ssc = new StreamingContext(sparkConf, Seconds(1))

    val host = "localhost"
    val port = 9999

    val lines = ssc.socketTextStream(host, port, StorageLevel.MEMORY_AND_
    DISK_SER)
    val words = lines.flatMap(_.split(" "))
    val wordCounts = words.map(x => (x, 1)).reduceByKey(_ + _)

    wordCounts.print()

    ssc.start()
    ssc.awaitTermination()
  }
}
```

There are a few important steps when putting together a DStream application. The entry point to a DStream application is the StreamingContext. One of the required inputs is the batch interval, which defines Spark's time duration to batch up a set of incoming data into an RDD for processing. It also represents a trigger point for when Spark should execute streaming application computation logic. For example, if the batch interval is 3 seconds, Spark batches all the data that arrive within that 3-second interval. After that interval elapses, it turns that batch of data into an RDD and processes it according to the processing logic you provide. Once a StreamingContext is created, the next step creates an instance DStream by defining an input source. The example defines the input source as a socket that reads lines of text. After this point, then you provide the processing logic for the newly created DStream. The processing logic in the preceding example is not complex. Once an RDD for a collection of lines is available after 1 second, Spark executes the logic of splitting each line into words, converting each word into a tuple of the word and a count of 1, and finally summing up all the count of the same word.

Finally, the counts are printed out on the console. Remember that a streaming application is a long-running application; therefore, it requires a signal to start receiving and processing the incoming stream of data. That signal is given by calling the StreamingContext start() function, which is usually done at the end of the file. The awaitTermination() function waits for the execution of the streaming application to stop and a mechanism to prevent the driver from exiting while your streaming application is running. In a typical program, once the last line of code is executed, it exits. However, a long-running streaming application needs to keep going once it starts and only ends when you explicitly stop it.

Like most first-generation stream processing engines, DStream has a few drawbacks.

- Native support for event time: For most stream processing applications, it is extremely important to extract insights or aggregations based on the event time. DStream, unfortunately, doesn't provide native support for this need.

- Separate APIs for batch and stream processing: Spark developers need to learn different APIs to build batch and streaming process applications. This isn't a fault of DStream, because Structured APIs were not available when DStream was invented.

Spark Structured Streaming

Structured Streaming is Spark's second-generation streaming engine. It was designed to be much faster, more scalable, and more fault-tolerant and address the shortcomings in the first-generation streaming engine. It was designed for developers to build end-to-end streaming applications to react to data in real-time using a simple programming model, which is built on top of the optimized and solid foundation of the Spark SQL engine. One distinguishing aspect of Structured Streaming is that it provides a unique and easy way for Spark users and developers to reason about building streaming applications.

Building production-grade streaming applications requires overcoming many challenges, and with that in mind, the Structured Streaming engine was designed to help deal with these challenges.

- Handling end-to-end reliability and guaranteeing correctness

- Ability to perform a complex transformation on various kinds of incoming data

- Processing of data based on event time and dealing with out-of-order data easily

- Integrating with a variety kind of data sources and data sinks

The following sections cover various aspects of the Structured Streaming engine and its support to deal with these challenges.

Overview

There are two key ideas in Structured Streaming. The first one treats a streaming computation the way a batch computation is treated. This means treating the incoming data stream as an input table, and as a new set of data arrives, treating it as a new set of rows appended to the input table (see Figure 6-7).

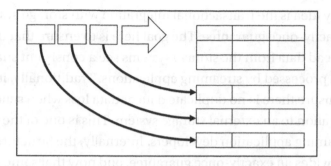

Figure 6-7. *Treating streaming data as a table being continuously updated*

Another way of thinking of a stream of incoming data is as nothing more than a table being continuously appended. This simple yet radical idea has many implications. One of them is leveraging the existing structured APIs for DataFrame and Dataset in Scala, Java, or Python to perform streaming computations. The Structured Streaming engine takes care of running them incrementally and continuously as new streaming data arrives. Figure 6-8 provides a visual comparison between performing batch and stream processing in Spark. The other implication is that the same Catalyst engine discussed in Chapter 5 optimizes the streaming computation logic expressed via structured APIs. The knowledge you gain from working with structured APIs is directly transferable to building streaming applications running on the Spark Structured Streaming engine. The only remaining parts to be learned are the ones that are specific to the stream processing domain, like event-time processing and maintaining state.

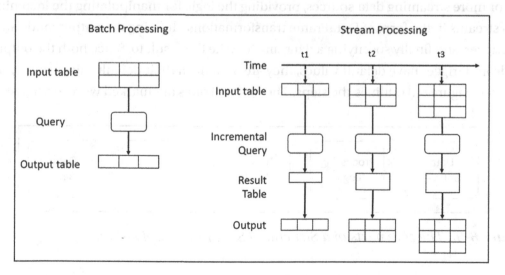

Figure 6-8. *Comparing batch processing and stream processing in Spark*

The second key idea is the transactional integration with storage systems to provide an end-to-end, exactly-once guarantee. The goal here is to ensure that the serving applications that read data from the storage systems see a consistent snapshot of the data that has been processed by streaming applications. Traditionally, it is a developers' responsibility to ensure there is no duplicate data or data loss when sending data from a streaming application to an external storage system. This is one of the pain points that was raised by streaming application developers. Internally, the Structured Streaming engine already provides an exactly-once guarantee, and now that same guarantee is extended to external storage systems, provided those systems support transactions.

Starting with Apache Spark 2.3, the Structured Streaming engine's processing model has been expanded to support a new model called *continuous processing*. The previous processing model was the micro-batching model, which is the default one. Given the nature of the micro-batching processing model, it is suitable for use cases that can tolerate end-to-end latency in the range of 100 milliseconds. For other use cases that need end-to-end latency as low as 1 millisecond, they should use the continuous processing model; however, it is in an experimental status as of Apache Spark 2.3 version. It has a few restrictions in terms of what streaming computations are supported.

Core Concepts

This section covers a set of core concepts you need to understand before building a streaming application. The main parts of a streaming application consist of specifying one or more streaming data sources, providing the logic for manipulating the incoming data streams in the form of DataFrame transformations, defining the output mode and the trigger, and finally specifying a data sink to write the result to. Since both the output mode and trigger have default values, they are optional if their default values meet your use case. Figure 6-9 outlines the steps. The optional ones are marked with an asterisk.

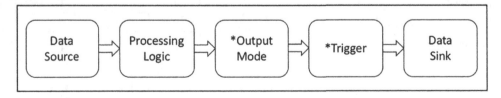

Figure 6-9. *The core parts of a Structured Streaming application*

Each of these concepts is described in detail in the following sections.

Data Sources

Let's start with data sources. With batching processing, the data source is a static dataset that resides on some storage system like the local file system, HDFS, or S3. The data sources in Structured Streaming are quite different. They generate the data continuously, and the rate can vary over time. Structured Streaming provides native support for the following sources.

- **Kafka source**: require Apache Kafka with version 0.10 or higher. This is the most popular data source in a production environment. Working with this data source requires a fundamental understanding of how Kafka works. Connecting to and reading data from a Kafka topic requires a specific set of settings that must be provided. Please refer to Kafka Integration Guide on the Spark website (`https://spark.apache.org/docs/latest/structured-streaming-kafka-integration.html`) for more information.

- **File source**: Files located on either local file system, HDFS or S3. As new files are dropped into a directory, this data source picks them up for processing. Commonly used file formats are supported, such as text, CSV, JSON, ORC, and Parquet. See the DataStreamReader interface for an up-to-date list of supported file formats. A good practice when working with this data source is to make the input files are completely written, then move them into the path of this data source.

- **Socket source**: This is for testing purposes only. It reads UTF8 data from a socket listening on a certain host and port.

- **Rate source**: This is for testing and benchmark purposes only. This source can be configured to generate several events per second, where each event consists of a timestamp and a monotonically increased value. This is the easiest source to work with while learning Structured Streaming.

One important property a data source needs to provide for Structured Streaming to deliver an end-to-end exactly-once guarantee is a marker. It can rewind to that location when reprocessing is needed. For example, Kafka data source provides a Kafka offset to track the read position of a partition of a topic. This property determines whether a particular data source is fault-tolerant or not. Table 6-1 describes some of the options for each of the out-of-the-box data sources.

Table 6-1. *Out-of-the-Box Data Sources*

Name	Fault-Tolerant	Configurations
File	Yes	`path`: Path to the input directory `maxFilesPerTrigger`: Maximum number of new files to read per trigger `latestFirst`: Whether to process the latest files (in terms of modification time) or not.
Socket	No	The following are required `host`: host to connect to `port`: port to connect to
Rate	Yes	`rowsPerSecond`: Number of rows to generate per second `rampUpTime`: Ramp up time in seconds before reaching `rowsPerSecond` `numPartitions`: number of partitions
Kafka	Yes	`kafka.bootstrap.servers`: A comma-separated list of `host:port` of Kafka brokers `subscribe`: a comma-separated list of topics Please refer to the Kafka Integration Guide on the Spark website for more information.

Apache Spark 2.3 introduced the Data Source V2 API, which is an officially supported set of interfaces for Spark developers to develop custom data sources that can easily integrate with Structured Streaming. With this well-defined set of APIs, the number of custom Structured Streaming sources dramatically increase.

Output Modes

Output modes are a way to tell Structured Streaming how the output data should be written to a sink. This concept is unique to stream processing in Spark. There are three options.

- **Append mode**: This is the default mode if the output mode is not specified. In this mode, only the new rows appended to the resulting table are sent to the specified output sink.

- **Complete mode**: The entire resulting table is written to the output sink.

- **Update mode**: Only the updated rows in the resulting table are written to the output sink. This means unchanged rows are not written out.

The semantics of the various Output modes take some time to get used to because there are a few dimensions to them. Given the three options, it is natural to wonder under what circumstances you would use one output mode versus the other ones. Hopefully, it makes more sense when you go through a few examples.

Trigger Types

The trigger is another important concept to understand. The Structured Streaming engine uses the trigger information to determine when to execute the provided streaming computation logic on the newly discovered streaming data. Table 6-2 describes the different trigger types.

Table 6-2. *Trigger Types*

Type	Description
Not specified (default)	For this default type, Spark uses the micro-batch mode and processes the next batch of data as soon as the previous batch of data has completed processing.
Fixed interval	For this type, Spark uses the micro-batch mode and processes the batch of data based on the user-provided interval. If the processing of the previous batch of data takes longer than the interval, then the next batch of data is processed immediately after the previous one is completed. In other words, Spark does not wait until the next interval boundary.
One-time	This trigger type is meant to be used for one time processing of the available batch of data, and Spark immediately stop the streaming application once the processing is completed. This trigger type is useful when the data volume is extremely low, and therefore it is more cost effective to spin up a cluster and process the data only a few times a day.
Continuous	Spark executes your streaming application logic using the new low latency and continuous processing mode.

Data Sinks

Data sinks are at the opposite end of the data sources. They are meant for storing the output of streaming applications. It is important to understand which sinks can support which output mode and whether they are fault-tolerant. A short description of each sink is provided here, and the various options for each sink are outlined in Table 6-3.

- **Kafka sink**: require Apache Kafka with version 0.10 or higher. There is a specific set of settings to connect to a Kafka cluster. Please refer to the Kafka Integration Guide on the Spark website for more information.

- **File sink**: This is a destination on a file system, HDFS, or S3. Commonly used file formats are supported, such as text, CSV, JSON, ORC, and Parquet. See the `DataStreamReader` interface for an up-to-date list of supported file formats.

- **Foreach sink**: This is meant for running an arbitrary computation on the rows in the output.

- **Console sink**: This is for testing and debugging purposes only and when working with low volume data. The output is printed out to the console on every trigger.

- **Memory sink**: This is for testing and debugging purposes only when working with low-volume data. It uses the memory of the driver to store the output.

Table 6-3. *Out-of-the-Box Data Sinks*

Name	Supported Output Modes	Fault Tolerant	Configurations
File	Append	Yes	path: path to the input directory All the popular file formats are supported. See DataFrameWriter for more information.
Foreach	Append, Update, Complete	Depends	This is a very flexible sink, and it is implementation-specific. See the following details.
Console	Append, Update, Complete	No	numRows: number of rows to print every trigger. Default is 20 rows truncate: whether to truncate if each row is too long. Default is true.
Memory	Append, Complete	No	N/A
Kafka	Append, Update, Complete	Yes	kafka.bootstrap.servers: a comma-separated list of host:port of Kafka brokers topic: a Kafka topic to write data to. Please refer to Kafka Integration Guide on Spark's website for more information.

One important property a data sink must support for Structured Streaming to deliver an end-to-end and exactly-once guarantee is idempotent for handling reprocessing. In other words, it must be able to handle multiple writes (that occur at different times) of the same data such that the outcome is the same as if there was only a single write. The multiple writes are a result of reprocessing data during a failure scenario.

The next section uses examples to demonstrate how the various pieces fit together when developing a Spark Structured Streaming application.

Watermarking

Watermarking is a commonly used technique in stream processing engines to deal with data that arrives at a much later time than other data created at about the same time. Late data presents challenges to stream processing engines when the streaming computation logic requires maintaining some state. Examples of this scenario are when there are aggregations or joining going on. Streaming application developers can specify a threshold to let the Structured Streaming engine know how late the data is expected to be in event time. With this information, the Structured Streaming engine can decide whether a piece of late data is processed or discarded.

More importantly, Structured Streaming uses the specified threshold to determine when the old state can be discarded. Without this information, Structured Streaming needs to maintain all the states indefinitely, and this causes out-of-memory issues for streaming applications. Any production Structured Streaming applications that perform aggregations or joining need to specify a watermark. This is an important concept, and more information about this topic is discussed and illustrated in later sections.

Structured Streaming Applications

This section walks through a Spark Structured Streaming example application to see how concepts are mapped into code. The following example is about processing a small set of mobile action events from a file data source. Each event consists of three fields.

- **id**: Represents the unique id of a phone. In the provided sample dataset, the phone ID is something like phone1, phone2, phone3.

- **action**: Represents an action taken by a user. Possible values of the action are open and close

- **ts**: Represents the timestamp when the action was taken by a user. This is the event time.

The mobile event data is split into three JSON files, which are available in the chapter6/data/mobile directory. To simulate the data streaming behavior, the JSON files are copied into the input folder in a certain order, and then the output is examined to validate your understanding.

Let's explore the mobile event data by using DataFrames to read the data (see Listing 6-2).

Listing 6-2. Reading in Mobile Data and Printing Its Schema

```
val mobileDataDF = spark.read.json("<path>/chapter6/data/mobile")

mobileDataDF.printSchema
 |-- action: string (nullable = true)
 |-- id: string (nullable = true)
 |-- ts: string (nullable = true)

file1.json
{"id":"phone1","action":"open","ts":"2018-03-02T10:02:33"}
{"id":"phone2","action":"open","ts":"2018-03-02T10:03:35"}
{"id":"phone3","action":"open","ts":"2018-03-02T10:03:50"}
{"id":"phone1","action":"close","ts":"2018-03-02T10:04:35"}

file2.json
{"id":"phone3","action":"close","ts":"2018-03-02T10:07:35"}
{"id":"phone4","action":"open","ts":"2018-03-02T10:07:50"}

file3.json
{"id":"phone2","action":"close","ts":"2018-03-02T10:04:50"}
{"id":"phone5","action":"open","ts":"2018-03-02T10:10:50"}
```

By default, Structured Streaming requires a schema when reading data from a file-based data source. This makes sense because it is not possible to infer the schema of the incoming streaming data when the directory is empty. However, if you want it to infer the schema, you can set the configuration spark.sql.streaming.schemaInference to true. In this example, you explicitly create a schema. Listing 6-3 contains a snippet of code for creating the schema for the mobile event data.

Listing 6-3. Create a Schema for Mobile Event Data

```
import org.apache.spark.sql.types._
import org.apache.spark.sql.functions._

val mobileDataSchema = new StructType()
                          .add("id", StringType, false)
                          .add("action", StringType, false)
                          .add("ts", TimestampType, false)
```

Let's start with a simple use case for processing the mobile event data. Our goal is to generate a count of each action type using a fixed window length with a ten-second. The three lines of code in Listing 6-4 help to achieve this goal. The first line illustrates using a file-based data source by using the DataStreamReader class to read data from a directory. The expected data format is in JSON, and the schema consists of three columns, defined in Listing 6-3. The returned object of the first line is an instance of DataFrame class. Unlike the DataFrame covered in Chapter 4, this DataFrame is a streaming DataFrame. You can simply confirm this by calling the isStreaming function, and the returned value should be true. The streaming computation logic in this simple application is expressed in the second line, which performs the group by transformation using the action column and a fixed window based on the ts column. The fixed window in the group by transformation is based on the timestamp embedded inside the mobile event data. The third line is important because it defines the output mode and data sink. Most importantly, it tells the Structured Streaming engine to start incrementally running the streaming computation logic expressed in the second line. To go into more detail, the third line of code uses the DataFrameWriter instance of the actionCountDF DataFrame to specify the console as the data sink, meaning the output is printed out to a console for you to examine. It then defines the output mode as "complete" so you can see all the records in the result table. Finally, it invokes the start() function of the DataStreamWriter class to start the execution, which means the data source starts processing files that are dropped into the /<path>/chapter6/data/input directory. Another important thing to note is that the start function returns an instance of a StreamingQuery class, representing a handle to a query that is continuously executing in the background as new data arrives. You can use the mobileConsoleSQ streaming query to examine the status and progress of the computation in your streaming application.

244

Before you type in the lines of code in Listing 6-4, make sure the input folder is empty.

Listing 6-4. Generate a Count Per Action Type in a 10-Second Sliding Window

```
// create a streaming DataFrame from reading data file in the specified
directory
val mobileSSDF = spark.readStream.schema(mobileDataSchema)
                    .json("/<path>/chapter6/data/input")

mobileSSDF.isStreaming


// perform a group by using event time of column ts and fixed window of 10 mins
val actionCountDF = mobileSSDF.groupBy(window($"ts",
                                "10 minutes"), $"action").count

// start the streaming query and write the output to console
val mobileConsoleSQ = actionCountDF.writeStream
                .format("console").option("truncate", "false")
                .outputMode("complete")
                .start()
```

The start() function in Listing 6-4 triggers the Spark Structured Streaming engine to start watching the input folder and start processing the data once it sees new files in that folder. After copying the file1.json file from the chapter6/data/mobile directory to the chapter6/data/input directory, the output console displays the output similar to the lines in Listing 6-5.

The output indicates there is only one window from 10:00 to 10:10, and within this window, there are one close action and three open actions, which should match the four lines of events in files1.json. Now repeat the same process with file2.json, and the output should match Listing 6-6. The file2.json data file contains one event with open action and another with close action and both fall in the same window. Therefore, the counts are updated to two close and four open respectively for action type.

Listing 6-5. Output from Processing file1.json

```
-------------------------------------------
Batch: 0
-------------------------------------------
+----------------------------------------------+--------+------+
|                                        window|  action| count|
+----------------------------------------------+--------+------+
|     [2018-03-02 10:00:00, 2018-03-02 10:10:00]|  close |     1|
|     [2018-03-02 10:00:00, 2018-03-02 10:10:00]|  open  |     3|
+----------------------------------------------+--------+------+
```

Listing 6-6. Output from Processing file2.json

```
-------------------------------------------
Batch: 1
-------------------------------------------
+---------------------------  -------------+--------+-------+
|                                    window|  action|  count|
+------------------------------------------+--------+-------+
|   [2018-03-02 10:00:00, 2018-03-02 10:10:00]|  close |     2|
|   [2018-03-02 10:00:00, 2018-03-02 10:10:00]|  open  |     4|
+------------------------------------------+--------+-------+
```

At this point, let's invoke a few functions of the query stream `mobileConsoleSQ` (an instance of `StreamingQuery` class) to examine the status and progress. The `status()` function tells you what's going on at the current state of the query stream, which can be either in wait mode or in the middle of processing the current batch of events. The `lastProgress()` function provides some metrics about the processing of the last batch of events, including processing rates, latencies, and so on. Listing 6-7 contains the sample output from both of these functions.

Listing 6-7. Output from Calling status() and lastProgress() Functions

```
scala> mobileConsoleSQ.status
res14: org.apache.spark.sql.streaming.StreamingQueryStatus =
{
  "message" : "Waiting for data to arrive",
```

```
  "isDataAvailable" : false,
  "isTriggerActive" : false
}

scala> mobileConsoleSQ.lastProgress
res17: org.apache.spark.sql.streaming.StreamingQueryProgress =
{
  "id" : "2200bc3f-077c-4f6f-af54-8043f50f719c",
  "runId" : "0ed4894c-1c76-4072-8252-264fe98cb856",
  "name" : null,
  "timestamp" : "2018-03-18T18:18:12.877Z",
  "batchId" : 2,
  "numInputRows" : 0,
  "inputRowsPerSecond" : 0.0,
  "processedRowsPerSecond" : 0.0,
  "durationMs" : {
    "getOffset" : 1,
    "triggerExecution" : 1
  },
  "stateOperators" : [ {
    "numRowsTotal" : 2,
    "numRowsUpdated" : 0,
    "memoryUsedBytes" : 17927
  } ],
  "sources" : [ {
    "description" : "FileStreamSource[file:<path>/chapter6/data/input]",
    "startOffset" : {
      "logOffset" : 1
    },
    "endOffset" : {
      "logOffset" : 1
    },
    "numInputRows" : 0,
    "inputRowsPerSecond" : 0.0,...
```

Let's finish processing the last file of the mobile event data. It's the same as `file2.json`. After `file3.json` is copied to the input directory, the output should look like Listing 6-8. File `file3.json` contains one close action that belongs to the first window and an open action that falls into a new window from 10:10 to 10:20. In total, there are eight actions. Seven of those fall into the first window, and one action falls into the second window.

Listing 6-8. Output from Processing file3.json

```
-------------------------------------------
Batch: 2
-------------------------------------------
+--------------------------------------------------+--------+-------+
|                                            window|  action|  count|
+--------------------------------------------------+--------+-------+
|    [2018-03-02 10:00:00, 2018-03-02 10:10:00]|   close|      3|
|    [2018-03-02 10:00:00, 2018-03-02 10:10:00]|    open|      4|
|    [2018-03-02 10:10:00, 2018-03-02 10:20:00]|    open|      1|
+--------------------------------------------------+--------+-------+
```

In a production and long-running streaming application, it is required to call the `StreamingQuery.awaitTermination()` function. It is a blocking call to prevent the main thread process from exiting and enable the streaming query to continuously run and process new data as they arrive into the data source. This function fails when the streaming query fails due to some foreseen errors.

While learning Structured Streaming, you may want to stop the streaming query to change the output mode, trigger, or other configurations. You can use the `StreamingQuery.stop()` function to stop the data source from receiving new data and stop the continuous execution of logic in a streaming query. Listing 6-9 shows examples of managing streaming queries.

Listing 6-9. Managing Streaming Query

```
// this is blocking call
mobileSQ.awaitTermination()

// stop a streaming query
mobileSQ.stop
```

```
// another way to stop all streaming queries in a Spark application
for(qs <- spark.streams.active) {
    println(s"Stop streaming query: ${qs.name} - active:
            ${qs.isActive}")
    if (qs.isActive) {
      qs.stop
    }
}
```

Streaming DataFrame Operations

Listing 6-9 shows that once a data source is configured and defined, the
DataStreamReader returns an instance of a DataFrame—the same one you are
familiar with from Chapter 3 and Chapter 4. This means you can use most operations
and Spark SQL functions to express your application's streaming computation
logic. It is important to note that not all operations in DataFrame are supported in
a streaming DataFrame. This is because some of them are not applicable in stream
processing, where the data is unbounded. Examples of such operations include
limit, distinct, cube, and sort.

Selection, Project, Aggregation Operations

One of the selling points of Structured Streaming is a set of unified APIs for batch
processing and stream processing in Spark. With a streaming DataFrame, it is
feasible to apply any of the select and filter transformations to it and any of the
Spark SQL functions that operate on individual columns. In addition, the basic
aggregations and the advanced analytics functions covered in Chapter 4 are also
available to a streaming DataFrame. A streaming DataFrame can be registered as a
temporary view and then apply SQL queries on it. Listing 6-10 provides an example
of filtering and applying Spark SQL functions on top of the mobileSSDF DataFrame in
Listing 6-4.

Listing 6-10. Apply Filtering and Spark SQL Functions on a Streaming DataFrame

```
import org.apache.spark.sql.functions._
val cleanMobileSSDF = mobileSSDF.filter($"action" === "open"
                                || $"action" === "close")
                        .select($"id", upper($"action"), $"ts")

// create a view to apply SQL queries on
cleanMobileSSDF.createOrReplaceTempView("clean_mobile")
spark.sql("select count(*) from clean_mobile")
```

It is important to note the following DataFrame transformations are not supported yet in a streaming DataFrame either because they are too complex to maintain state or because of the unbounded nature of streaming data.

- Multiple aggregations or a chain of aggregations on a streaming DataFrame

- Limit and take N rows

- Distinct transformation (There is a way to deduplicate data using a unique identifier, however.)

- Sorting on a streaming DataFrame without any aggregation (sorting is supported after some form of aggregation, however.)

Any attempt to use one of the unsupported operations results in an AnalysisException exception. You see a message that states something like, "operation XYZ is not supported with streaming DataFrames/Datasets".

Join Operations

One of the coolest things you can do with a streaming DataFrame is to join it with either a static DataFrame or another streaming DataFrame. Joining is a complex operation and the tricky part is that not all of the data for a streaming DataFrame is available at the time of joining. Therefore, the result of a join is generated incrementally at each trigger point, similar to how the result of an aggregation is generated.

Starting with Spark version 2.3, Structured Streaming supports joining two streaming DataFrames. Given the unbounded nature of a streaming DataFrame, Structured Streaming must maintain the past data of both streaming DataFrames to match with any future, yet-to-be-received data. To avoid the explosion of the streaming state that Structured Streaming must maintain, a watermark can be optionally provided for both streaming DataFrames, and a constraint on event-time must be defined in the join condition. Let's go through an IoT use case of joining two data sensor–related data streams of a data center. The first one contains the temperature reading of the various locations in a data center. The second one contains the load information of each computer in the same data center. The joint condition of these two streams is the location. Listing 6-11 contains code about providing watermarks and a constraint on the event-time in the join condition.

Listing 6-11. Joining Two Streaming DataFrames

```
import org.apache.spark.sql.functions.expr

// the specific streaming data source information is not important in this
   example
val tempDataDF = spark.readStream. ...
val loadDataDF = spark.readStream. ...

val tempDataWatermarkDF = tempDataDF.withWaterMark("temp_taken_time",
"1 hour")
val loadDataWatermarkDF = loadDataDF.withWaterMark("load_taken_time",
"2 hours")

// join on the location id as well as the event time constraint
tempWithLoadDataDF = tempDataWatermarkDF.join(loadDataWatermarkDF,
   expr(""" temp_location_id = load_location_id AND
           load_taken_time >= temp_taken_time AND
           load_taken_time <= temp_taken_time + interval 1 hour
      """)
)
```

There are more restrictions on the outer joins when joining a static DataFrame and a streaming DataFrame and two streaming DataFrames. Table 6-4 provides some information.

Table 6-4. *Some Details About Joining Streaming DataFrame*

Left Side+Right Side	Join Type	Note
Static+Streaming	Inner	Supported
Static+Streaming	Left Outer	Not supported
Static+Streaming	Right Outer	Supported
Static+Streaming	Full Outer	Not supported
Streaming+Streaming	Inner	Supported
Streaming+Streaming	Left Outer	Conditionally supported. Must specify watermark on right side and time constraint
Streaming+Streaming	Right Outer	Conditionally supported. Must specify watermark on left and time constraint
Streaming+Streaming	Full Outer	Not supported

Working with Data Sources

The previous section described each of the built-in sources that Structured Streaming provides. This section goes into more detail and provides sample codes for working with them.

Both socket and rate data sources are designed for testing and learning purposes only, and they shouldn't be used in production.

Working with a Socket Data Source

The socket data source is easy to work with, and it only requires information about the host and port to connect to. Before starting a streaming query for the socket data source, it is important to start a socket server first using a network utility command-line utility like nc on macOS or netcat for Windows. In this example, the nc network utility is used, and you need to bring up two terminals. The first one is for starting up a socket server with port number 9999; the command is nc -lk 9999. The second one is for running the Spark shell with the code in Listing 6-12.

Listing 6-12. Reading Streaming Data from Socket Data Source

```
val socketDF = spark.readStream.format("socket")
                           .option("host", "localhost")
                           .option("port", "9999").load()

val words = socketDF.as[String].flatMap(_.split(" "))
val wordCounts = words.groupBy("value").count()

val query = wordCounts.writeStream.format("console")
                    .outputMode("complete")
                    .start()
```

Now go back to the first terminal, type **Spark is great**, and hit the Enter key. Then type Spark is awesome and hit the enter key. Hitting the enter key tells the Netcat server to send whatever was typed to the socket listener. If everything went well, there should be two output batches in the Spark shell console, as in Listing 6-13, and each one contains the count of each word. Since Structured Streaming maintains state across batches, it was able to update the count of the words Spark and is to 2.

Listing 6-13. Output of Socket Data Source in Spark-Shell Console

```
-------------------------------------------------
Batch: 0
-------------------------------------------------
+--------+-------+
|   value|  count|
+--------+-------+
|   great|      1|
|      is|      1|
|   Spark|      1|
+--------+-------+
```

```
------------------------------------------------
Batch: 1
------------------------------------------------
+------------+-------+
|       value|  count|
+------------+-------+
|       great|      1|
|          is|      2|
|     awesome|      1|
|       Spark|      2|
+------------+-------+
```

When you are done testing the Socket data source, feel free to stop the streaming query by calling the `stop` function, as shown in Listing 6-14.

Listing 6-14. Stop a Streaming Query of Socket Data Source

```
query.stop
```

Working with a Rate Data Source

Like the socket data source, the rate data source was designed for testing and learning purposes only. It supports a few options, and the key one is the number of rows to generate per second. If that number is high, then another optional configuration can be provided to specify the ramp-up time to get to the number of rows per second. Each piece of data the rate source produces contains two columns: the timestamp and the auto-increment value. Listing 6-15 contains the code for printing out the data from the rate data source and what the first batch looks like in the console.

Listing 6-15. Working with Rate Data Source

```
// configure it to generate 10 rows per second
val rateSourceDF = spark.readStream.format("rate")
                                   .option("rowsPerSecond","10")
                                   .load()
```

```
val rateQuery = rateSourceDF.writeStream
                        .outputMode("update")
                        .format("console")
                        .option("truncate", "false")
                        .start()
-------------------------------------------
Batch: 1
-------------------------------------------
+------------------------------+-------+
|                     timestamp|  value|
+------------------------------+-------+
|           2018-03-19 10:30:21.952|   0   |
|           2018-03-19 10:30:22.052|   1   |
|           2018-03-19 10:30:22.152|   2   |
|           2018-03-19 10:30:22.252|   3   |
|           2018-03-19 10:30:22.352|   4   |
|           2018-03-19 10:30:22.452|   5   |
|           2018-03-19 10:30:22.552|   6   |
|           2018-03-19 10:30:22.652|   7   |
|           2018-03-19 10:30:22.752|   8   |
|           2018-03-19 10:30:22.852|   9   |
+------------------------------+-------+
```

One interesting thing to note is the number in the value column is guaranteed to be consecutive across all the partitions. Listing 6-16 illustrates what the output looks like with three partitions.

Listing 6-16. the Output of Rate Data Source with the Partition ID

```
import org.apache.spark.sql.functions._

// with 3 partitions
val rateSourceDF2 = spark.readStream.format("rate")
                        .option("rowsPerSecond","10")
                        .option("numPartitions",3).load()
```

```
// add partition id column to examine
val rateWithPartitionDF =
  rateSourceDF2.withColumn("partition_id", spark_partition_id())

val rateWithPartitionQuery = rateWithPartitionDF.writeStream
                                  .outputMode("update")
                                  .format("console")
                                  .option("truncate", "false")
                                  .start()

// output of batch one
-------------------------------------------
Batch: 1
-------------------------------------------

+-----------------------------+--------+-------------+
|                    timestamp|   value| partition_id|
+-----------------------------+--------+-------------+
|       2018-03-24 08:46:43.412|   0    |      0      |
|       2018-03-24 08:46:43.512|   1    |      0      |
|       2018-03-24 08:46:43.612|   2    |      0      |
|       2018-03-24 08:46:43.712|   3    |      1      |
|       2018-03-24 08:46:43.812|   4    |      1      |
|       2018-03-24 08:46:43.912|   5    |      1      |
|       2018-03-24 08:46:44.012|   6    |      2      |
|       2018-03-24 08:46:44.112|   7    |      2      |
|       2018-03-24 08:46:44.212|   8    |      2      |
|       2018-03-24 08:46:44.312|   9    |      2      |
+-----------------------------+--------+-------------+
```

The output shows the ten rows are spread across three partitions, and the values are consecutive as if they were generated for a single partition. If you are curious about the implementation of this data source, then check out https://github.com/apache/spark/blob/master/sql/core/src/main/scala/org/apache/spark/sql/execution/streaming/RateSourceProvider.scala.

Working with a File Data Source

The file data source is simplest to understand and work with. Let's say there is a need to process new files that are periodically copied into a directory. This is the perfect data source for this use case. Out of the box, it supports all the commonly used file formats, including text, CSV, JSON, ORC, and Parquet. For a complete list of supported file formats, please consult the DataStreamReader interface. Among the four options that the file data source supports, the only required one is the input directory to read files from.

As new files are copied into a specified directory, the file data source picks them up to process. It is possible to configure this data source to selectively pick up only a fixed number of new files for processing. The option to specify the number of files is called maxFilesPerTrigger.

Listing 6-17 provides an example of reading JSON mobile data events from a directory and using the same schema defined in Listing 6-3. Another interesting and optional option that the file data source supports is processing the latest files before the older files. It uses the timestamp of a file to determine which file is newer. The default behavior is to process files from oldest to latest. This option is useful when there is a large backlog of files to process, and you want to process the new files first.

Listing 6-17. Working with File Data Source

```
val mobileSSDF = spark.readStream.schema(mobileDataSchema)
                      .json("<directory name>")

// if you want to specify maxFilesPerTrigger
val mobileSSDF = spark.readStream.schema(mobileDataSchema)
                      .option("maxFilesPerTrigger", 5)
                      .json("<directory name>")

// if you want to process new files first
val mobileSSDF = spark.readStream.schema(mobileDataSchema)
                      .option("latestFirst", "true")
                      .json("<directory name>")
```

Working with a Kafka Data Source

Most production streaming applications process streaming data from Kafka data sources. To be effective at working with this data source, you need to have basic knowledge about working with Kafka. At a high level, this data source acts as a Kafka consumer, so the information it needs is very similar to what a typical Kafka consumer needs. There are two required pieces of information and a handful of optional ones.

The two required ones are a list of Kafka servers to connect to and information about one or more topics to consume the data from. In terms of flexibility and support a variety of needs, it supports three different ways of specifying this information. You just need to pick the one that best suits your use case. Table 6-5 contains information about the two required options.

Table 6-5. *Required Options for Kafka Data Source*

Option	Value	Description
kafka.bootstrap. servers	host1:port1, host2:port2	This is a comma-separated list of Kafka broker servers. Consult your Kafka administrators for hostname and port number to use
subscribe	topic1,topic2	This is a comma-separated list of topic names for this data source to read data from.
subscribePattern	topic.*	This is a regex pattern to express which topics to read data from. It is a little bit more flexible than the subscribe option.
assign	{ topic1: [1,2], topic2: [3,4] }	With this option, you can specify the specific list of partitions of the topics to read data from. This information must be provided in JSON format.

After the required options are specified, you can optionally specify the options in Table 6-5, which contains only a subset of the commonly used ones. For a complete list of optional options, please consult the Structured Streaming and Kafka Integration Guide. The reason these options are optional is that they have default values.

The startingOffsets and endingOffsets options are a way for you to have fine-grain control of processing data in Kafka from a specific point in a particular partition of a particular topic. This flexibility is extremely useful in scenarios where reprocessing is needed due to a failure, bugs introduced in a new version of the software, or when retraining a machine learning model. The ability to reprocess data in Kafka is one of the reasons that Kafka is very popular in the world of big data processing. It may be obvious, but the startingOffsets is used by the Kafka data source to figure out where to start reading the data from in Kafka. and therefore, once the processing is going, this option is no longer used. The endingOffsets is used by the Kafka data source to figure out when to stop reading the data from Kafka. For example, if you want your streaming application to read the latest data from Kafka and continue with processing new incoming data, the latest are the startingOffsets and endingOffsets values.

Table 6-6. *Optional Options for Kafka Data Source*

Option	Default Value	Value	Description
startingOffsets	latest	earliest, latest JSON string of starting offset for each topic, i.e., { "topic1": { "0":45, "1": -1}, "topic2": { "0":-2} }	earliest means the beginning of a topic. latest means whatever the latest data is in a topic. When using the JSON string format, −2 represents the earliest offset in a specific partition, and −1 represents the latest offset in a specific partition
endingOffsets	latest	latest JSON string, i.e. { "topic1": { "0":45, "1": -1}, "topic2": { "0":-2} }	latest means the latest data in a topic. When using the JSON string format, −1 represents the latest offset in a specific partition. Naturally, −2 is not applicable for this option.
maxOffsets PerTrigger	none	Long. i.e., 500	This option is a rate limit mechanism to control the number of records to process per trigger interval. If a value is specified, it represents the total number of records across all the partitions, not per partition.

By default, the Kafka data source is not included in the Apache Spark binary available at `https://spark.apache.org/downloads.html`. If you want to use the Kafka data source from the Spark shell, it is important to start the Spark shell with an extra option to download and include the right jar file. The deployment section of Structured Streaming and Kafka integration documentation (`https://spark.apache.org/docs/latest/structured-streaming-kafka-integration.html`) provides the information about the extra option. It looks something like in Listing 6-18.

Listing 6-18. Start Spark Shell with Kafka Data Source Jar File

```
./bin/spark-shell --packages org.apache.spark:spark-sql-kafka-0-10_2.11:2.3.0

// if the above package is not provided, the following problem will be
  encountered

java.lang.ClassNotFoundException: Failed to find data source: kafka. Please
find packages at http://spark.apache.org/third-party-projects.html
  at org.apache.spark.sql.execution.datasources.DataSource$.
  lookupDataSource(DataSource.scala:635)
  at org.apache.spark.sql.streaming.DataStreamReader.load(DataStreamReader.
  scala:159)
```

Let's start with a simple example of processing the data from the beginning of a Kafka topic called `pageviews` and continue processing new data as they arrive in Kafka. Listing 6-19 shows the code.

Listing 6-19. Kafka Data Source Example

```
import org.apache.spark.sql.functions._

val pvDF = spark.readStream.format("kafka")
            .option("kafka.bootstrap.servers","localhost:9092")
            .option("subscribe", "pageviews")
            .option("startingOffsets", "earliest")
            .load()

pvDF.printSchema
 |-- key: binary (nullable = true)
 |-- value: binary (nullable = true)
```

```
|-- topic: string (nullable = true)
|-- partition: integer (nullable = true)
|-- offset: long (nullable = true)
|-- timestamp: timestamp (nullable = true)
|-- timestampType: integer (nullable = true)
```

One unique thing about the Kafka data source is that the streaming DataFrame it returns has a fixed schema, which looks like Listing 6-19. The value column contains the actual content of each Kafka message, and the column type is binary. Kafka doesn't care about the content of each message, and therefore it treats it as a binary blob. The remaining columns in the schema contain the metadata of each message. If the content of the messages was serialized in some binary format at the time of sending to Kafka, then you need a way to deserialize it using either Spark SQL functions or a UDF before those messages can be processed in Spark.

In Listing 6-20, the content is a string, so you simply need to cast it to a String type. For demonstration purposes, Listing 6-20 performs the casting of the value column and selects a few metadata-related columns to display.

Listing 6-20. Casting Message Content To String Type

```
val pvValueDF = pvDF.selectExpr("partition","offset",
                "CAST(key AS STRING)", "CAST(value AS STRING)")
                    .as[(String, Long, String, String)]
```

The examples in Listing 6-21 contain a few variations of specifying Kafka topic, partition, and offset to read Kafka messages.

Listing 6-21. Various Examples of Specifying Kafka Topic, Partition and Offset

```
// reading from multiple topics with default startingOffsets and
   endingOffsets
val kafkaDF = spark.readStream.format("kafka")
   .option("kafka.bootstrap.servers","server1:9092,server2:9092")
   .option("subscribe", "topic1,topic2")
   .load()
```

```
// reading from multiple topics using subscribePattern
val kafkaDF = spark.readStream.format("kafka")
    .option("kafka.bootstrap.servers","server1:9092,server2:9092")
    .option("subscribePattern", "topic*")
    .load()

// reading from a particular offset of a partition using JSON format
// the triple quotes format in Scala is used to escape double quote in
    JSON string
Val kafkaDF = spark.readStream.format("kafka")
        .option("kafka.bootstrap.servers","localhost:9092")
        .option("subscribe", "topic1,topic2")
        .option("startingOffsets", """ {"topic1": {"0":51} } """)
        .load()
```

Working with a Custom Data Source

Starting with Spark 2.3 version, the Data Source APIs V2 was introduced to address the issues in V1 and provide a set of new APIs that are clean, extensible, and easy to work with. The Data Source APIs V2 is available in Scala only.

This section is meant to provide a quick overview of the interfaces and main APIs involved in building a custom data source using Data Source APIs V2. Before doing so, it is a good idea to study the implementation of a few of the built-in data sources, such as the RateSourceProvider.scala, RateSourceProviderV2.scala, and KafkaSourceProvider.scala classes.

All custom data sources must implement a marker interface called DataSourceV2, and then it can decide whether to implement interface ContinuousReadSupport or MicroBatchReadSupport or both. For example, KafkaSourceProvider.scala implements both interfaces because it allows users to choose which processing mode to use based on a use case. Each of the two interfaces acts as a factory method for creating an instance of ContinuousReader or MicroBatchReader, respectively. The bulk of the custom data source implementation is in implementing the APIs defined in these two interfaces.

I've implemented a fun and non-fault-tolerant data source that reads wiki edits from the Wikipedia IRC server. It is fairly easy to use Spark Structured Streaming to analyze the wiki edits of various Wikipedia sites. See the README.md in the GitHub repository (https://github.com/beginning-spark/beginning-apache-spark-3/tree/master/ chapter6/custom-data-source) for more information. To use this custom data source in Spark shell, the first step is to download the streaming_sources-assembly-0.0.1.jar file from the GitHub repository. Listing 6-22 describes the remaining steps

Listing 6-22. Analyzing Wiki Edits with a Custom Data Source

```
// start up spark-shell with streaming_sources-assembly-0.0.1.jar
bin/spark-shell --jars <path>/streaming_sources-assembly-0.0.1.jar

// once spark-shell is successfully started

// define the data source provider name
val provideClassName = "org.structured_streaming_sources.wikedit.
WikiEditSourceV2"

// use custom data and subscribe to English Wikipedia edit channel
val wikiEditDF = spark.readStream.format(provideClassName).
option("channel", "#en.wikipedia").load()

// examine the schema of wikiEditDF streaming DataFrame

wikiEditDF.printSchema

 |-- timestamp: timestamp (nullable = true)
 |-- channel: string (nullable = true)
 |-- title: string (nullable = true)
 |-- diffUrl: string (nullable = true)
 |-- user: string (nullable = true)
 |-- byteDiff: integer (nullable = true)
 |-- summary: string (nullable = true)

// select only a few columns for analysis
val wikiEditSmallDF = wikiEditDF.select("timestamp", "user", "channel",
"title")

// start streaming query and write out the wiki edits to console
val wikiEditQS = wikiEditSmallDF.writeStream.format("console").
option("truncate", "false").start()
```

```
// wait for a few seconds for data to come in and the result might looking
like below
+------------------------+------------+-------------+---------------------------+
|           timestamp    |    user    |   channel   |            title          |
+------------------------+------------+-------------+---------------------------+
| 2018-03-24 15:36:39.409| 6.62.103.211| #en.wikipedia| Thomas J.R. Hughes       |
| 2018-03-24 15:36:39.412|  .92.206.108| #en.wikipedia| List of international schools|
+------------------------+------------+-------------+---------------------------+

// to stop the query stream
wikiEditQS.stop
```

Notice the custom data source name is a fully qualified class name of the data source provider. It is not short like the built-in data sources because those already registered their short name in a file called org.apache.spark.sql.sources.DataSourceRegister.

Working with Data Sinks

The last step in a streaming application involves writing the computation result to some storage system or sending it to some downstream system for consumption. Structured Streaming provides five built-in sinks, and three of them are for production usage, and the remaining ones are for testing purposes. The following sections go into detail on each one and provide sample codes for working with them.

Working with a File Data Sink

The file data sink is a simple to understand and work with. The only required option you need to provide is the output directory. Since the file data sink is fault-tolerant, Structured Streaming requires a checkpoint location to write the progress information and other metadata to help with the recovery when there was a failure.

The example in Listing 6-23 configures the rate data source to generate ten rows per second, send the generated rows to two partitions, and write the data in JSON format to the specified directory.

Listing 6-23. Write Data from Rate Data Source To File Sink

```
val rateSourceDF = spark.readStream.format("rate")
                        .option("rowsPerSecond","10")
                        .option("numPartitions","2")
                        .load()

val rateSQ = rateSourceDF.writeStream.outputMode("append")
                        .format("json")
                        .option("path", "/tmp/output")
                        .option("checkpointLocation", "/tmp/ss/cp")
                        .start()

// use the line below to stop the writing the data
rateSQ.stop
```

Since the number of partitions was configured as two, Structured Streaming writes the output out to two files to the specified output folder at each trigger point. So, if you examine the output folder, you see files with names that start with either part-00000 and part-00001. The rate data source was configured with ten rows per second, and there are two partitions. Therefore, each output contains five rows, as shown in Listing 6-24.

Listing 6-24. the Content of Each Output File

```
{"timestamp":"2018-03-24T17:42:08.182-07:00","value":205}
{"timestamp":"2018-03-24T17:42:08.282-07:00","value":206}
{"timestamp":"2018-03-24T17:42:08.382-07:00","value":207}
{"timestamp":"2018-03-24T17:42:08.482-07:00","value":208}
{"timestamp":"2018-03-24T17:42:08.582-07:00","value":209}
```

Working with a Kafka Data Sink

In Structured Streaming, writing the data of a streaming DataFrame to Kafka data sink is simpler than reading data from Kafka data source. The Kafka data sink can be configured with the four options listed in Table 6-7. Three of the options are required. The important options to understand are the key and value related to the Kafka message structure. The unit of data in Kafka is a message, which essentially is a key-value pair. The role of the value is to hold the actual content of a Kafka message.

As far as Kafka is concerned, the value is just a collection of bytes. Kafka considers key as metadata, and it is saved along with the value in the Kafka message. When a message is sent to Kafka, and a key is provided, Kafka utilizes it as a routing mechanism to determine which partition a particular Kafka message should be sent to by hashing the key and performing a modulo on the number of partitions a particular topic has. This implies that all messages with the same key are routed to the same partition. If a key is not provided, Kafka can't guarantee which partition that message is sent to, and Kafka employs a round-robin algorithm to balance the messages between partitions.

Table 6-7. *Options for Kafka Data Sink*

Option	Value	Description
kafka.bootstrap. servers	host1:port1, host2:port2	This is a comma-separated list of Kafka broker servers. Consult your Kafka administrators for hostname and port number to use
topic	topic1	This is a topic name
key	A string or binary	This key determines which partition a Kafka message should be sent to. All Kafka messages with the same key go to the same partition. This is an optional option.
value	A string or binary	This is the content of a message. To Kafka, it is simply just an array of bytes.

There are two ways to provide a topic name. The first way is to provide it in the configuration of a Kafka data sink and the second way is by defining a column in the streaming DataFrame called `topic`. The value of that column is used as the topic name.

If a streaming DataFrame has a column called `key`, that column value is used as the message key. Since the key is an optional piece of metadata, it is not required to have this column in the streaming DataFrame. On the other hand, the value must be provided, and Kafka data sink expects a column named `value` in the streaming DataFrame.

Listing 6-25 provides an example of setting a rate data source and then writes the data out to a Kafka topic called `rates`. If you are planning to use the Spark shell to try the code, include the necessary argument described to include the `org.apache. spark:spark-sql-kafka-0-10_2.11:2.3.0` jar file and its dependencies.

Note   The simplest way to get started with Kafka is to download the Confluent Platform package and follow its Getting Started guide. More information is available at `https://docs.confluent.io/current/getting-started.html`. Once the download is completed, uncompress the compressed tar file into a directory. To start up the servers (Zookeeper, Kafka Broker, Schema Registry), use the ./bin/confluent start command line. Each of those servers listens on a specific port. All the command line tools are available in the bin directory, and almost all of them require the host and port for either Zookeeper or Kafka Broker. Before running the code in Listing 6-21, make sure to create a topic called *rates*. The command to do that is bin/kafka-topics --create --zookeeper localhost:2181 --replication-factor 1 --partitions 2 --topic rates. To list active topics, use this command: ./bin/kafka-topics --zookeeper localhost:2181 --list.

Listing 6-25. Write Data from Rate Data Source To File Sink

```
import org.apache.spark.sql.functions._

// setting up the rate data source with 10 rows per second and use two
    partitions
val ratesSinkDF = spark.readStream.format("rate")
                    .option("rowsPerSecond","10")
                    .option("numPartitions","2")
                    .load()

// transform the ratesSinkDF to create a column called "key" and "value"
    column
// the value column contains a JSON string that contains two fields:
timestamp and value
val ratesSinkDF = ratesSinkDF.select(
            $"value".cast("string") as "key",
            to_json(struct("timestamp","value")) as "value")

// setup a streaming query to write data to Kafka using topic "rates"
val rateSinkSQ = ratesSinkDF.writeStream
                        .outputMode("append")
```

```
                              .format("kafka")
                              .option("kafka.bootstrap.servers",
                                        "localhost:9092")
                              .option("topic","rates")
                              .option("checkpointLocation",
                                        "/Users/hluu/tmp/ss/cp")
                              .start()

// it doesn't take long to write a lot of messages to Kafka, so after a few
second, feel free to stop the
// rateSinkSQL
rateSinkSQ.stop
```

To read the data back from the rates topic in Kafka, use the sample code listed in Listing 6-21 and substitute the appropriate value for options such as kafka. bootstrap.servers and topic name. The data in the rates Kafka topic look something like Listing 6-22.

Listing 6-26. Sample of Data from Kafka

```
+---------+---------+---------+-----------------------------------------------------------+
|partition| offset|    key  | value                                                       |
+---------+---------+---------+-----------------------------------------------------------+
|    1    |  9350   |   583249| {"timestamp":"2018-03-25T09:53:52.582-07:00","value":583249}|
|    1    |  9351   |   583250| {"timestamp":"2018-03-25T09:53:52.682-07:00","value":583250}|
|    1    |  9352   |   583251| {"timestamp":"2018-03-25T09:53:52.782-07:00","value":583251}|
|    1    |   9353  |   583256| {"timestamp":"2018-03-25T09:53:53.282-07:00","value":583256}|
|    1    |   9354  |   583261| {"timestamp":"2018-03-25T09:53:53.782-07:00","value":583261}|
|    1    |   9355  |   583266| {"timestamp":"2018-03-25T09:53:54.282-07:00","value":583266}|
|    1    |   9356  |   583267| {"timestamp":"2018-03-25T09:53:54.382-07:00","value":583267}|
|    1    |   9357  |   583274| {"timestamp":"2018-03-25T09:53:55.082-07:00","value":583274}|
|    1    |   9358  |   583275| {"timestamp":"2018-03-25T09:53:55.182-07:00","value":583275}|
|    1    |   9359  |   583276| {"timestamp":"2018-03-25T09:53:55.282-07:00","value":583276}|
+---------+---------+---------+-----------------------------------------------------------+
```

Working with a foreach Data Sink

Compared to the other built-in data sinks that Structured Streaming provides, the foreach data sink is interesting because it provides complete flexibility in terms of how the data should be written, when to write out the data, and where to write the data to. It was designed to be an extensible as well as a pluggable data sink. This flexibility and extensibility come with a responsibility because you are responsible for the logic of writing out the data. In a nutshell, you need to provide an implementation of the ForeachWriter abstract class, which consists of three methods: open, process, and close. They get called whenever there is a list of output rows after a trigger. Working with this data sink requires some intimate details about how Spark works.

- An instance of the ForeachWriter abstract class implementation is created on the driver side, and it is sent to the executors in your Spark cluster for execution. This has two implications. First, the implementation of ForeachWriter must be serializable; otherwise, an instance of it can't be shipped across the network to the executors. Second, if there are any initializations during the creation of the implementation, they happen on the driver side. For example, if you want to create a database or socket connection, that should not happen during the class initialization but rather somewhere else.

- The number of partitions in a streaming DataFrame determines how many instances of the ForeachWriter implementation are created. This is very similar to the behavior of the Dataset.foreachPartition method.

- The three methods defined in the ForeachWriter abstract class are invoked on the executor's side.

- The open method is the best place to perform initializations like opening a database connection or socket connect. However, it is called each time data is written out; therefore, that logic must be intelligent and efficient.

- The open method signature has two input parameters: partition id and version. Boolean is the return type. The combination of these two parameters uniquely represents a set of rows that needs to be written out. The value of the version is a monotonically increasing id that increases with every trigger. Based on the value of the partition id and version parameters, the open method needs to decide whether it needs to write out the sequence of rows or not and return the appropriate boolean value for a Structured Streaming engine.

- If the open method returns true, then the process method is called for each row of the output of a trigger.

- The close method is guaranteed to be called. If there was an error during the call to the process method, that error is passed to the close method. The intention for calling the close method is to give you a chance to clean up any necessary state that was created during the open or process method invocation. The only time the close method is not called is when the JVM of the executor crashes or the open method throws a throwable exception.

In short, this data sink provides the ultimate flexibility in writing out the data of a streaming DataFrame. Listing 6-27 contains a very simple implementation of the ForeachWriter abstract class by writing the data from the rate data source out to the console.

Listing 6-27. Sample Code for Working with Foreach Data Sink

```
// define an implementation of the ForeachWriter abstract class
import org.apache.spark.sql.{ForeachWriter,Row}

class ConsoleWriter(private var pId:Long = 0, private var ver:Long = 0)
extends ForeachWriter[Row] {
    def open(partitionId: Long, version: Long): Boolean = {
       pId = partitionId
       ver = version
       println(s"open => ($partitionId, $version)")
       true
    }

    def process(row: Row) = {
      println(s"writing => $row")
    }
```

```
  def close(errorOrNull: Throwable): Unit = {
    println(s"close => ($pId, $ver)")
  }
}

// setup the Rate data source
val ratesSourceDF = spark.readStream.format("rate")
                        .option("rowsPerSecond","10")
                        .option("numPartitions","2")
                        .load()

// setup the Foreach data sink
val rateSQ = ratesSourceDF.writeStream.foreach(new ConsoleWriter).start()

// sample output from the console
open => (1, 1)
writing => [2018-03-25 13:03:41.867,5]
writing => [2018-03-25 13:03:41.367,0]
writing => [2018-03-25 13:03:41.967,6]
writing => [2018-03-25 13:03:41.467,1]
writing => [2018-03-25 13:03:42.067,7]
writing => [2018-03-25 13:03:41.567,2]
writing => [2018-03-25 13:03:42.167,8]
writing => [2018-03-25 13:03:41.667,3]
writing => [2018-03-25 13:03:42.267,9]
close => (1, 1)

// to close the rateSQ streaming query
rateSQ.stop
```

Working with a Console Data Sink

This console data sink is easy to work with. It does exactly what it sounds. It is not a fault-tolerant data sink. It is designed to be used for debugging purposes or while learning Structured Streaming. The two options it provides are the number of rows to display and whether to truncate the output if too long. Each option has a default value, as shown in Table 6-8. The underlying implementation of this data sink uses the same logic as in the DataFrame.show method to display the data in a streaming DataFrame.

Table 6-8. *Options for Console Data Sink*

Option	Default Value	Description
numRows	20	The number of rows to print to console
truncate	true	Whether to truncate with the content of each column is longer than 20 characters

Listing 6-28 shows the console data sink in action and provides a value for each of the two options.

Listing 6-28. Sample Code for Working with Console Data Sink

```
// setting up a data source
val ratesDF = spark.readStream.format("rate")
                    .option("rowsPerSecond","10")
                    .option("numPartitions","2")
                    .load()

Val ratesSQ = ratesDF.writeStream.outputMode("append")
                    .format("console")
                    .option("truncate",false)
                    .option("numRows",50)
                    .start()
```

Working with a Memory Data Sink

Like the console data sink, the memory data sink is very easy to understand and work with. It is so easy because it has no options that you need to provide. It is not a fault-tolerant data sink. It is designed to be used for debugging purposes or while learning Structured Streaming. The data it collects is sent to the driver and stored on the driver as an in-memory table. In other words, the amount of data you can send to the memory data sink is bounded by the amount of memory the driver JVM has. While setting up this data sink, you can specify a query name as an argument to `DataStreamWriter. queryName` function. Then you can issue SQL queries against the in-memory table. Unlike the console data sink, once the data is sent to the in-memory table, you can further analyze or process the data using pretty much all the features available in the Spark SQL component. If the amount of data is large and wouldn't fit into memory, the next best option is to use the file data sink to write the data out in Parquet format.

The sample code in Listing 6-29 writes the data from the rate data source into an in-memory table, and issues Spark SQL queries against the in-memory table.

Listing 6-29. Sample Code for Working with the Memory Data Sink

```
val ratesDF  = spark.readStream.format("rate")
                   .option("rowsPerSecond","10")
                   .option("numPartitions","2")
                       .load()
// write data out to Memory data sink with in-memory table name as "rates"
val ratesSQ = ratesDF.writeStream.outputMode("append")
                     .format("memory")
                     .queryName("rates")
                     .start()

// you issue SQL queries against the "rates" in-memory table
spark.sql("select * from rates").show(10,false)
+---------------------------------+-------+
|            timestamp            | value|
+---------------------------------+-------+
|            2018-03-25 14:02:59.461|  0   |
|            2018-03-25 14:02:59.561|  1   |
|            2018-03-25 14:02:59.661|  2   |
|            2018-03-25 14:02:59.761|  3   |
|            2018-03-25 14:02:59.861|  4   |
|            2018-03-25 14:02:59.961|  5   |
|            2018-03-25 14:03:00.061|  6   |
|            2018-03-25 14:03:00.161|  7   |
|            2018-03-25 14:03:00.261|  8   |
|            2018-03-25 14:03:00.361|  9   |
+---------------------------------+-------+
```

```
// count the number of rows in the "rates" in-memory table
spark.sql("select count(*) from rates").show
+-----------+
|   count(1)|
+-----------+
|        100|
+-----------+

// to stop the ratesSQ query stream
ratesSQ.stop
```

One thing to note is that the in-memory `rates` table still be around after the ratesSQ streaming query has stopped. However, once a new streaming query is started with the same name, then the data from in-memory is truncated

Before concluding this section, it is important to understand which outputs are supported by each type of data sink. Table 6-9 is a quick summary for reference. The output modes are covered in the next section.

Table 6-9. *Data Sinks and Their Support Output Modes*

Sink	Supported Output Modes	Notes
File	Append	Support writing out new rows only and no update
Kafka	Append,Update,Complete	
Foreach	Append,Update,Complete	Depends on the ForeachWriter implementation
Console	Append,Update,Complete	
Memory	Append,Complete	Doesn't support in-place updates

Output Modes

The "Output Modes" section described each of the output modes. This section provides more information about them and ways to understand which output mode is applicable for which streaming query type.

There are two types of streaming queries. The first type is called the *stateless type,* and it performs only basic transformations on the incoming streaming data and then writes out the data to one or more data sinks. The second type is the *stateful type,* which requires maintaining some state between trigger points, whether that is done implicitly or explicitly. The stateful type usually performs aggregations or uses the Structured Streaming APIs like `mapGroupsWithState` or `flatMapGroupsWithState` to maintain some arbitrary state needed for a particular use case; for example, maintaining user session data.

Let's start with the simple, stateless streaming query type. A typical use case for this kind of streaming query is the real-time streaming ETL. It continuously reads the incoming streaming data like page view events produced by online services to capture which pages are being viewed by their users. In this kind of use case, it usually performs the following.

- Filtering, transforming, and cleaning. Real-world data is messy. The structure may not be well suited for repeated analysis.

- Converting to a more efficient storage format. Text, CVS, and JSON are human-readable file formats but are inefficient for repeated analysis, especially if the data volume is large such as hundreds of terabytes. More efficient binary formats like ORC, Parquet, or Avro are more suitable to reduce data size and improve analysis speed.

- Partitioning data by certain columns. While writing the data out to a data sink, it is possible to partition the data based on the value of commonly used columns in queries to speed up the repeated analysis from various teams in the organization.

As you can see, the tasks don't require a streaming query to maintain any kind of state before writing the data out to a data sink. As new data comes in, it is cleaned, transformed, and possibly restructured, and finally written out. `append` is the only applicable output mode for this stateless streaming type. The `complete` output mode is not applicable because that requires Structured Streaming to maintain all the previous data, which may be too large. The `update` output mode is not applicable because only new data is being written out. However, when this output mode is used for a stateless streaming query, Structured Streaming recognizes this and treats it the same as the `append` output mode. The cool thing is when an inappropriate output mode is used for a streaming query, the Structured Streaming engine lets you know. Listing 6-30 shows what happens when an inappropriate output mode is used.

Listing 6-30. Using "Complete" Output Mode with a Stateless Streaming Query

```
val ratesDF  = spark.readStream.format("rate")
                    .option("rowsPerSecond","10")
                    .option("numPartitions","2")
                    .load()

// simple transformation
val oddEvenDF = ratesDF.withColumn("even_odd",
                                    $"value" % 2 === 0)

// write out to Console data sink using complete output mode
val ratesSQ = oddEvenDF.writeStream.outputMode("complete")
                    .format("console")
                    .option("truncate",false)
                    .option("numRows",50)
                    .start()

// An exception from Structured Streaming during the analysis phase
org.apache.spark.sql.AnalysisException: Complete output mode not supported
when there are no streaming aggregations on streaming DataFrames/Datasets;
```

Now let's move on to the second query type. When a steaming query performs an aggregation via a groupBy transformation, the state of that aggregation is maintained implicitly by the Structured Streaming engine. As more data comes in, the result of the aggregation on new data is updated into the result table. At each trigger point, the updated data or all the data in the result table is written to a data sink, depending on the output mode. This implies that using the append output mode is inappropriate because that violates the semantics of that output mode, which specifies that only new rows appended to the result table are sent to the specified output sink. In other words, only the complete and update output modes are appropriate for stateful query types. The output of a streaming query using the complete output mode is always equal to or more than the output of the same streaming query using the update output mode. Listing 6-31 contains the code to illustrate the difference in the output.

Listing 6-31. the Output Differences Between Update and Complete Mode

```
// import statements
import org.apache.spark.sql.types._
import org.apache.spark.sql.functions._

val schema = new StructType().add("id", StringType, false)
                             .add("action", StringType, false)
                             .add("ts", TimestampType, false)

val mobileDF = spark.readStream.schema(schema)
                 .json("<path>/chapter6/data/input")

val actionCountDF = mobileDF.groupBy($"action").count

val completeModeSQ = actionCountDF.writeStream.format("console")
                                  .option("truncate", "false")
                                  .outputMode("complete")
                                  .start()

val updateModeSQ = actionCountDF.writeStream.format("console")
                                .option("truncate", "false")
                                .outputMode("complete").start()

// at this point copy file1.json, file2.json, file3.json and newaction.json
from
// mobile directory to the input directory

// the output of the streaming query with complete mode is below
-----------------------------------------------
Batch: 3
-----------------------------------------------
+--------+-------+
| action|  count|
+--------+-------+
|  close | 3     |
|  swipe | 1     |
|  crash | 1     |
|  open  | 5     |
+--------+-------+
```

```
// the output of the streaming query with update mode is below
-------------------------------------------
Batch: 3
-------------------------------------------
+-------+--------+
| action|   count|
+-------+--------+
| swipe |   1    |
| crash |   1    |
+-------+--------+
```

The output of the streaming query with the Complete output mode contains all the action types in the result table. The output of the streaming query with the update output mode contains only the actions in the newaction.json file that the result table hasn't seen before.

Again, if an inappropriate output mode is used for a stateful query type, the Structured Streaming engine lets you know, as shown in Listing 6-32.

Listing 6-32. Using an Inappropriate "Append" Output Mode with a Stateful Streaming Query

```
// use an inappropriate output for stateful streaming query, see exception
below
val actionCountSQ = actionCountDF.writeStream.format("console")
                            .outputMode("append").start()
```

org.apache.spark.sql.AnalysisException: Append output mode not supported when there are streaming aggregations on streaming DataFrames/DataSets without watermark;

There is an exception to this logic. All the output modes are applicable if a watermark is provided to the stateful streaming query with aggregation. The semantics of the Append output are not violated anymore because the Structured Streaming engine drops the old aggregation state data that is older than the specified watermark, which means new rows can be added to the result table once the watermark is crossed.

Undoubtedly, the output mode is one of the most complicated concepts in Structured Streaming because multiple dimensions come together to determine which output modes are appropriate to use. The Structured Streaming programming guide provides a compatibility matrix, which is at `https://spark.apache.org/docs/latest/structured-streaming-programming-guide.html#output-modes`.

Triggers

The trigger setting determines when the Structured Streaming engine runs the streaming computation logic expressed in a streaming query, which includes all the transformation logic and writing out the data to the data sink. Another way of thinking about it is that the trigger setting controls when the data is written out to a data sink and which processing mode to use. Starting with Spark version 2.3, a new processing mode called *continuous* was introduced.

The "Trigger Types" section describes the supported types in Structured Streaming. This section goes into more detail and provides a sample code for specifying the different trigger types.

All the stream query examples have used the default trigger type, which is used when a trigger type is not specified. This default trigger type chooses the micro-batch mode as the processing mode, and the logic in the streaming query is executed not based on time but as soon as the previous batch of data has completed processing. This implies there is less predictability in terms of how often the data is written out.

If a little more predictability is desired, then the fixed interval trigger can be specified to tell Structured Streaming to execute the streaming query logic at a certain time interval based on the user-provided value, for example, every 30 seconds. In terms of processing mode, this trigger type uses the micro-batch one. The interval can be specified in a string format or as a Scala `Duration` or Java `TimeUnit`. Listing 6-33 contains examples for using the fixed interval trigger.

Listing 6-33. Examples of Using Fixed Interval Trigger Type

```
import org.apache.spark.sql.streaming.Trigger

// setting up with 3 rows per second
val ratesDF  = spark.readStream.format("rate")
                    .option("rowsPerSecond","3")
                    .option("numPartitions","2")
                    .load()

// trigger the streaming query execution every 3 seconds and write out to
   console
val ratesSQ = ratesDF.writeStream.outputMode("append")
                    .format("console")
                    .option("numRows",50)
                    .option("truncate",false)
                    .trigger(Trigger.ProcessingTime("3 seconds"))
                    .start()

// you should expect to see about 9 rows in every 3 seconds
+--------------------------------+-------+
|            timestamp           | value|
+--------------------------------+-------+
|         2018-03-26 07:14:11.176|    0 |
|         2018-03-26 07:14:11.509|    1 |
|         2018-03-26 07:14:11.843|    2 |
|         2018-03-26 07:14:12.176|    3 |
|         2018-03-26 07:14:12.509|    4 |
|         2018-03-26 07:14:12.843|    5 |
|         2018-03-26 07:14:13.176|    6 |
|         2018-03-26 07:14:13.509|    7 |
|         2018-03-26 07:14:13.843|    8 |
+--------------------------------+-------+

// specifying the interval using Scala Duration type
import scala.concurrent.duration._
```

```
val ratesSQ = ratesDF.writeStream.outputMode("append")
                    .format("console")
                    .option("numRows",50)
                    .option("truncate",false)
                    .trigger(Trigger.ProcessingTime(3.seconds))
                    .start()
```

The Fixed interval trigger provides the best effort. It can't guarantee the execution of a streaming query always happens exactly at the specified internal. There are two reasons for this. The first one is when there is no incoming data, then there is nothing to process, and therefore nothing is written out the data sink. The second reason is that when the previous batch's processing time exceeds the specified interval, the next execution of a streaming query starts as soon as the processing completes. In other words, it does not wait for the next interval boundary.

The one-time trigger does what it sounds like. It executes the logic in a streaming query in a micro-batch mode and writes out the data to a data sink one time, and then the processing stops. It may sound silly for this trigger type to exist; however, it is very useful in both development and production environments. While in the development phase, the streaming computation logic is usually developed in an iterative manner, and in each iteration, you want to test the logic. This trigger type simplifies the develop-test iteration a bit. For a production environment, this trigger type is suitable for use cases where the volume of incoming streaming data is low. Therefore, it is only necessary to run the streaming application a few times a day. Instead of launching a Spark cluster and leaving it running throughout the day, the frequency of launching a Spark and executing the stream processing logic one time or multiple times per day is based on the desired processing frequency of your particular use cases. It is quite simple to specify this one-time trigger type. Listing 6-34 shows how to do that.

Listing 6-34. Example of Using One-Time Trigger Type

```
import org.apache.spark.sql.streaming.Trigger

val mobileSQ =  mobileDF.writeStream.outputMode("append")
                      .format("console")
                      .trigger(Trigger.Once())
                      .start()
```

The *continuous trigger type* is a new, exciting, and experimental processing mode was introduced in Spark version 2.3 to address the use cases that need end-to-end streaming millisecond latency. In this new processing mode, Structured Streaming launches long-running tasks to continuously read, process, and write data to a data sink. This implies the incoming data is processed and written out to the data sink as soon as they arrive in the data source, and the end-to-end latency is within a few milliseconds. In addition, an asynchronous checkpoint mechanism was introduced to record the streaming query progress efficiently to not interrupt the long-running tasks from providing consistent millisecond-level latencies. A good use case to leverage this trigger type is credit card fraudulent transaction detection. At a high level, the Structured Streaming engine figures out which processing mode to use based on the trigger type, which is depicted in Figure 6-10.

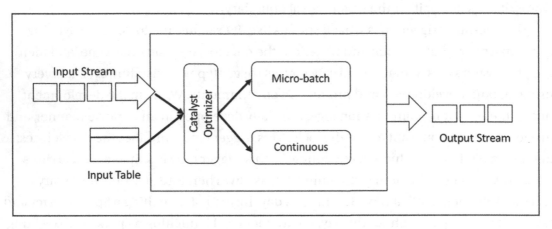

Figure 6-10. *Structured Streaming supports two different processing modes*

As of Spark version 2.4, only the projection and selections operations are allowed in the continuous processing mode, such as select, where, map, flatmap, and filter. In this processing mode, all Spark SQL functions are supported except aggregation functions.

To use the continuous processing mode for a streaming query, you must specify a continuous trigger with the desired checkpoint interval like in Listing 6-35.

Listing 6-35. Examples of Specifying a Continuous Trigger Type

```
import org.apache.spark.sql.streaming.Trigger

// setting a Rate data source with two partitions
val ratesDF  = spark.readStream.format("rate")
                    .option("numPartitions","2").load()

// write out the data to console and using continuous trigger with 2 second
interval for writing out progress
val rateSQ = ratesDF.writeStream.format("console")
                    .trigger(Trigger.Continuous("2 second"))
                    .start()

// sample output from console
+-------------------------+-------+
|                timestamp| value|
+-------------------------+-------+
|     2018-03-26 21:43:...|      0|
|     2018-03-26 21:43:...|      2|
|     2018-03-26 21:43:...|      4|
|     2018-03-26 21:43:...|      6|
|     2018-03-26 21:43:...|      1|
|     2018-03-26 21:43:...|      3|
|     2018-03-26 21:43:...|      5|
|     2018-03-26 21:43:...|      7|
+-------------------------+-------+
```

The ratesDF streaming DataFrame was set up with two partitions; therefore, Structured Streaming launched two running tasks in the continuous processing mode. That is why the output shows all the even numbers appearing together and all the odd numbers appearing together.

Summary

Structured Streaming is the second-generation stream processing engine of Apache Spark. It provides an easy way to build and reason about fault-tolerant and scalable streaming applications. This chapter covers a lot of ground, including the core concepts in the stream processing domain and the key parts of Structured Streaming.

- Stream processing is an exciting domain that can help solve many new and interesting use cases in the era of big data.

- Building production streaming data applications is much more challenging than building batch data processing applications because of the nature of the unbounded data and the unpredictability of the data arrival rate and out-of-order data.

- To be effective at building streaming data applications, you must be comfortable with the three core concepts in the stream processing domain. They are data delivery semantics, the notion of time, and windowing.

- Stream processing engines have drastically and dramatically matured in the last few years, and now there are many options to choose from. The popular ones are Apache Flink, Apache Samza, Apache Kafka, and Apache Spark.

- Spark DStream is the first-generation stream processing engine of Apache Spark. It was built on top of the RDD programming model.

- Structured Stream processing engine was designed for developers to build end-to-end streaming applications that can react to data in real-time using a simple programming model built on top of the optimized and solid foundation of the Spark SQL engine.

- The unique idea in Structured Streaming is to treat streaming data as an unbounded input table, and as new data arrives, it treats that as a new set of new rows to append to an unbounded table.

- The core components in streaming query are the data source, streaming operations, output mode, trigger, and data sink.

- Structured Streaming provides a set of built-in data sources as well as data sinks. The built-in data sources are File, Kafka, Socket, and Rate. The built-in data sinks are File, Kafka, Console, and Memory.

- Output mode determines how the data is output to a data sink. There are three options: Append, Update, and Complete.

- A trigger is a mechanism for a Structured Streaming engine to determine when to run the streaming computation. There are several options to choose from: micro-batch, fixed interval micro-batch, one-time micro-batch, and continuous. The last one is for use cases that need millisecond latency. It is in an experimental state as of Spark version 2.3.

CHAPTER 7

Advanced Spark Streaming

Chapter 6 introduced the core concepts in streaming processing, the Spark Structured Streaming processing engine's features, and the basic steps of developing a streaming application. Real-world streaming applications usually need to extract insights or patterns from the incoming real-time data at scale and feed that information into downstream applications to make business decisions or save that information in some storage system for further analysis or visualization purposes. Another aspect of real-world streaming applications is that they are continuously running to process real-time data as it comes in. Therefore, they must be resilient against failures.

The first half of this chapter covers event-time processing and stateful processing features in Structured Streaming and how they can help extract insights or patterns from incoming real-time data. The second half of this chapter explains the support that Structured Streaming provides to help streaming applications be fault-tolerant against failures and monitor their status and progress.

Event Time

The ability to process incoming streaming data based on the data creation time is a must-have feature for any serious streaming processing engine. This is important because to truly understand and accurately extract insights or patterns from streaming data. You need to process them based on when that data or those events happened, not when they are processed. Often, the event time processing is in the context of aggregation, which includes the event time and zero or more pieces of additional information in the event.

287

© Hien Luu 2021
H. Luu, *Beginning Apache Spark 3*, https://doi.org/10.1007/978-1-4842-7383-8_7

Let's take the example of the mobile action events described in Chapter 6. Instead of applying the aggregations over the action type, you can apply the aggregations over a time window, which could be a fixed or a sliding window type (described in Chapter 6). In addition, you can easily add the action type to the grouping key to further group the mobile action events by time bucket and action type.

The following example process the mobile data event; Listing 7-1 shows its schema. The ts column represents the time when an event was created, in other words, when a user opens or closes an application. The mobile event data is located in <path>/ chapter6/data/mobile directory, containing file1.json, file2.json, file3.json, and newaction.json. Listing 7-2 displays the content in each of those files.

Listing 7-1. Mobile Data Event Schema

```
mobileDataDF.printSchema

 |-- action: string (nullable = true)
 |-- id: string (nullable = true)
 |-- ts: timestamp (nullable = true)
```

Listing 7-2. Mobile Event Data in file1.json, file2.json, file3.json, newaction.json

```
// file1.json
{"id":"phone1","action":"open","ts":"2018-03-02T10:02:33"}
{"id":"phone2","action":"open","ts":"2018-03-02T10:03:35"}
{"id":"phone3","action":"open","ts":"2018-03-02T10:03:50"}
{"id":"phone1","action":"close","ts":"2018-03-02T10:04:35"}

// file2.json
{"id":"phone3","action":"close","ts":"2018-03-02T10:07:35"}
{"id":"phone4","action":"open","ts":"2018-03-02T10:07:50"}

// file3.json
{"id":"phone2","action":"close","ts":"2018-03-02T10:04:50"}
{"id":"phone5","action":"open","ts":"2018-03-02T10:10:50"}

// newaction.json
{"id":"phone2","action":"crash","ts":"2018-03-02T11:09:13"}
{"id":"phone5","action":"swipe","ts":"2018-03-02T11:17:29"}
```

Fixed Window Aggregation over an Event Time

A fixed window (a.k.a. a tumbling window) operation discretizes a stream of incoming data into nonoverlapping buckets based on window length. Each piece of incoming data is placed into one of the buckets based on its event time. Performing aggregations is just a matter of going through each bucket and applying the aggregation logic, whether a count or sum. Figure 7-1 illustrates the fixed window aggregation logic.

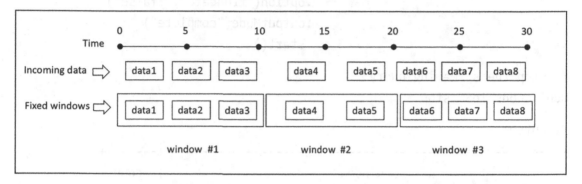

Figure 7-1. *Fixed window operation*

An example of fixed window aggregation is to perform a counting aggregation of the number of mobile events per each fixed window of ten minutes long. The window length is usually determined by the needs of a particular use case and the data volume. The aggregation result gives you high-level insights into the rate of mobile events that were generated per window. If you are interested in mobile usage throughout the day and by the hour, maybe the window length of 60 minutes is more appropriate. Listing 7-3 contains the code for performing the counting aggregation and the aggregation result. As expected, there are only ten mobile data events in all four files listed, and the total count in the output matches that number.

Listing 7-3. Process Mobile Event Data with a 10 Minute Window

```
import org.apache.spark.sql.types._
import org.apache.spark.sql.functions._

val mobileDataSchema = new StructType()
                    .add("id", StringType, false)
                    .add("action", StringType, false)
                    .add("ts", TimestampType, false)
```

289

```
val mobileSSDF = spark.readStream.schema(mobileDataSchema)
                    .json("<path>/chapter6/data/input")

val windowCountDF = mobileSSDF.groupBy(
                              window($"ts", "10 minutes"))
                          .count()

val mobileConsoleSQ = windowCountDF.writeStream.format("console")
                          .option("truncate", "false")
                          .outputMode("complete")
                          .start()

// stop the streaming query
mobileConsoleSQ.stop

// output
+------------------------------------------------+-------+
|                     window                     | count|
+------------------------------------------------+-------+
|       [2018-03-02 10:00:00, 2018-03-02 10:10:00]|      7|
|       [2018-03-02 10:10:00, 2018-03-02 10:20:00]|      1|
|       [2018-03-02 11:00:00, 2018-03-02 11:10:00]|      1|
|       [2018-03-02 11:10:00, 2018-03-02 11:20:00]|      1|
+------------------------------------------------+-------+

windowCountDF.printSchema

 |-- window: struct (nullable = false)
 |    |-- start: timestamp (nullable = true)
 |    |-- end: timestamp (nullable = true)
 |-- count: long (nullable = false)
```

When performing an aggregation with a window, the output window is a struct type, and it contains the start and end time.

In addition to specifying a window in the groupBy transformation, you can also specify additional columns from the event itself. Listing 7-4 performs the aggregation with a window length and the action. This gives additional insights into the count of each window and action type. It requires only a small change to the preceding example to accomplish this. Listing 7-4 contains only the lines that needed changes.

Listing 7-4. Process Mobile Event Data with a 10 Minute Window and Action Type

```
val windActDF= mobileSSDF.groupBy(
                    window($"ts", "10 minutes"), $"action").count

val windActDFSQ = windActDF.writeStream.format("console")
                           .option("truncate", "false")
                           .outputMode("complete")
                           .start()
// result
+-------------------------------------------------+--------+-------+
|                  window                         | action| count|
+-------------------------------------------------+--------+-------+
|    [2018-03-02 10:00:00, 2018-03-02 10:10:00]|  close |      3|
|    [2018-03-02 11:00:00, 2018-03-02 11:10:00]|  crash |      1|
|    [2018-03-02 11:10:00, 2018-03-02 11:20:00]|  swipe |      1|
|    [2018-03-02 10:00:00, 2018-03-02 10:10:00]|  open  |      4|
|    [2018-03-02 10:10:00, 2018-03-02 10:20:00]|  open  |      1|
+-------------------------------------------------+--------+-------+

// stop the query stream
windowActionCountSQ.stop()
```

Each line in this result table contains insights about the number of action types in each 10-minute window. If there are many crash actions in a certain window, that insight is useful if there was a release around that time frame.

Sliding Window Aggregation over Event Time

In addition to the fixed window type, there is another windowing type called *sliding window*. Defining a sliding window requires two pieces of information, the window length and a sliding interval, which is usually smaller than the window length. Given the aggregation computation is sliding over the incoming stream of data, the result is usually smoother than the result of fixed window type. Therefore, this windowing type is often used to compute moving averages. An important thing to note about a sliding window is that a piece of data can fall into more than one window because of the overlapping, as illustrated in Figure 7-2.

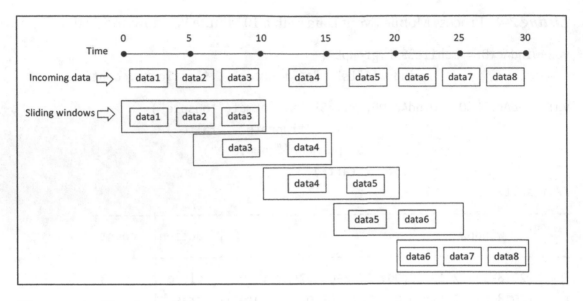

Figure 7-2. *Fixed window operation*

To illustrate the sliding window aggregation over the incoming data, you use small synthetic data about the temperature of computer racks in a data center. Imagine each computer rack emits its temperature at a certain interval. You want to generate a report about the average temperature among all computer racks and per rack over a window length of 10 minutes and a sliding interval of 5 minutes. This dataset is in the <path>/ chapter7/data/iot directory, which contains file1.json and file2.json. The temperature data is shown in Listing 7-5.

Listing 7-5. Temperature Data of Two Racks

```
// file1.json
{"rack":"rack1","temperature":99.5,"ts":"2017-06-02T08:01:01"}
{"rack":"rack1","temperature":100.5,"ts":"2017-06-02T08:06:02"}
{"rack":"rack1","temperature":101.0,"ts":"2017-06-02T08:11:03"}
{"rack":"rack1","temperature":102.0,"ts":"2017-06-02T08:16:04"}

// file2.json
{"rack":"rack2","temperature":99.5,"ts":"2017-06-02T08:01:02"}
{"rack":"rack2","temperature":105.5,"ts":"2017-06-02T08:06:04"}
{"rack":"rack2","temperature":104.0,"ts":"2017-06-02T08:11:06"}
{"rack":"rack2","temperature":108.0,"ts":"2017-06-02T08:16:08"}
```

Listing 7-6 first reads the temperature data, then performs a `groupBy` transformation on a sliding window over the `ts` column. For each sliding window, the `avg()` function is applied to the temperature column. To make it easy to inspect the output, it writes the data out to a memory data sink with a query name of `iot`. Then you can issue SQL queries against this temporary table.

Listing 7-6. Average Temperature of All the Computer Racks over a Sliding Window

```
import org.apache.spark.sql.types._
import org.apache.spark.sql.functions._

// define schema
val iotSchema = new StructType().add("rack", StringType, false)
                                .add("temperature",
                                      DoubleType, false)
                                .add("ts", TimestampType, false)

val iotSSDF = spark.readStream.schema(iotSchema)
                   .json("<path>/chapter7/data/iot")
// group by a sliding window and perform average on the temperature column
val iotAvgDF = iotSSDF.groupBy(window($"ts",
                                10 minutes", "5 minutes"))
                   .agg(avg("temperature") as "avg_temp")

// write the data out to memory sink with query name as iot
val iotMemorySQ = iotAvgDF.writeStream.format("memory")
                          .queryName("iot")
                          .outputMode("complete")
                          .start()

// display the data in the order of start time
spark.sql("select * from iot")
     .orderBy($"window.start").show(false)

// output
```

```
+----------------------------------------------------+-------------+
|                      window                         |    avg_temp|
+----------------------------------------------------+-------------+
|      [2017-06-02 07:55:00, 2017-06-02 08:05:00]|          99.5|
|      [2017-06-02 08:00:00, 2017-06-02 08:10:00]|         101.25|
|      [2017-06-02 08:05:00, 2017-06-02 08:15:00]|         102.75|
|      [2017-06-02 08:10:00, 2017-06-02 08:20:00]|         103.75|
|      [2017-06-02 08:15:00, 2017-06-02 08:25:00]|          105.0|
+----------------------------------------------------+-------------+
```

```
// stop the streaming query
iotMemorySQ.stop
```

This output shows five windows in the synthetic dataset. Notice the start time of each window is five minutes apart because of the length of the sliding interval you specified in the groupBy transformation. The temperature column indicates the average temperature is increasing, which is alarming. It is unclear whether the temperature of all the computer racks is increasing or only certain ones.

To help identify which computer racks, Listing 7-7 adds the rack column to the groupBy transformation, and it shows only the lines that are different from in Listing 7-6.

Listing 7-7. Average Temperature of Each Rack over a Sliding Window

```
// group by a sliding window and rack column
val iotAvgByRackDF = iotSSDF.groupBy(
                    window($"ts", "10 minutes", "5 minutes"),
                          $"rack")
                .agg(avg("temperature") as "avg_temp")

// write out to memory data sink with iot_rack query name
val iotByRackConsoleSQ = iotAvgByRackDF.writeStream
                                    .format("memory")
                                    .queryName("iot_rack")
                                    .outputMode("complete")
                                    .start()
```

```
spark.sql("select * from iot_rack").orderBy($"rack",
                $"window.start").show(false)
```

```
+-------------------------------------------+-------+------------+
|                                    window | rack  |   avg_temp|
+-------------------------------------------+-------+------------+
|[2017-06-02 07:55:00, 2017-06-02 08:05:00]| rack1|        99.5|
|[2017-06-02 08:00:00, 2017-06-02 08:10:00]| rack1|       100.0|
|[2017-06-02 08:05:00, 2017-06-02 08:15:00]| rack1|      100.75|
|[2017-06-02 08:10:00, 2017-06-02 08:20:00]| rack1|       101.5|
|[2017-06-02 08:15:00, 2017-06-02 08:25:00]| rack1|       102.0|
|[2017-06-02 07:55:00, 2017-06-02 08:05:00]| rack2|        99.5|
|[2017-06-02 08:00:00, 2017-06-02 08:10:00]| rack2|       102.5|
|[2017-06-02 08:05:00, 2017-06-02 08:15:00]| rack2|      104.75|
|[2017-06-02 08:10:00, 2017-06-02 08:20:00]| rack2|       106.0|
|[2017-06-02 08:15:00, 2017-06-02 08:25:00]| rack2|       108.0|
+-------------------------------------------+-------+------------+
```

```
// stop query stream
iotByRackConsoleSQ.stop()
```

The output table clearly shows the average temperature of rack 1 is below 103, and it is rack 2 that you should be concerned about.

Aggregation State

The previous examples of performing aggregations of over fixed or sliding windows with event time and additional columns show how easy it is to perform commonly used and complex streaming processing operations in Spark Structured Streaming. While it seems easy from the usage perspective, both the Structure Streaming engine and the Spark SQL engine work hard and cooperatively together to maintain the intermediate aggregation result in a fault-tolerant manner while executing the streaming aggregation. Any time an aggregation is performed on a streaming query, the intermediate aggregation state must be maintained. This state is maintained in key-value pairs structure, similar to a hash map, where the key is the group name and the value is the intermediate aggregation value. In the previous example of aggregation by a sliding window and rack ID, the key is the combined value of the start and end time of the window and the rack name, and the value is the average temperature.

The intermediate state is stored in an in-memory, versioned, key/value "state store" on the Spark executors. It is written to a write-ahead log, which should be configured to reside on a persistent storage system like HDFS. The state is read and updated at every trigger point in the in-memory "state store" and then written out to the write-ahead log. After a failure when a Spark Structured Streaming application is restarted, the state is restored from the write-ahead log and resumes from that point. This fault-tolerant state management incurs some resource and processing overheads in the Structure Streaming engine. The amount of overhead is proportional to the amount of state it needs to maintain. Therefore it is important to keep the amount of state in an acceptable size; in other words, the size of the state should not grow indefinitely.

Given the nature of sliding windows, the number of windows grows indefinitely. This implies that a sliding window aggregation requires the intermediate state to grow indefinitely unless there is a way to drop the old state that is no longer updated. This is accomplished using a technique called *watermarking*.

Watermarking: Limit State and Handle Late Data

Watermarking is a commonly used technique in streaming processing engines to deal with late data and limit the amount of state needed to maintain. Streaming data in the real world often arrive out of order or late because of network congestion, network disruption, or the data generator like mobile devices are not online. As a developer of real-time streaming applications, it is important to understand the tradeoff decision in dealing with the late data that arrives after a certain threshold. In other words, what is an acceptable amount of time you expect most of the data arrives relative to the others? Most likely, the answer to the preceding question is it depends on the use case. It might be that it is acceptable to drop the late data on the floor and ignore them.

From the perspective of Structured Streaming, a watermark is a moving threshold in event time that trails behind the maximum event time seen so far. As new data arrives, the maximum event time is updated, which causes the watermark to move. Figure 7-3 illustrates an example where the watermark is defined as ten minutes. The solid line represents the watermark line. It is trailing behind the maximum event timeline, which is represented by the dotted line. Each rectangular box represents a piece of data, and its event-time is immediately below the box. The piece of data with event-time 10:07 arrives a bit late, around 10:12; however, it still falls within the threshold between 10:03

and 10:13. Therefore it is processed as usual. The piece of data with event time 10:15 falls in the same category. The piece of data with event time 10:04 arrives late, around 10:22, which falls below the watermark line, and therefore it is ignored and not processed.

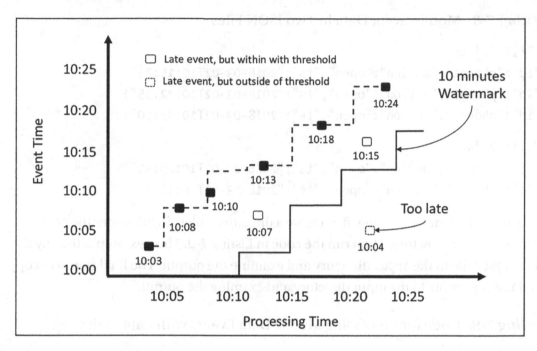

Figure 7-3. *Handling late date with watermark*

One of the biggest benefits of specifying the watermark is to enable the Structure Streaming engine to safely remove the aggregation state that is older than the watermark. Production streaming applications that perform aggregations should specify a watermark to avoid out-of-memory issues. Without a doubt, watermarking is an essential tool to deal with the messy part of real-time streaming data.

Structured Streaming makes it very easy to specify a watermark as a part of the streaming DataFrame. You just need to provide two pieces of data to the withWatermark API, the event time column, and the threshold, which can be in seconds, minutes, or hours. To demonstrate the watermark in action, you can work through a simple example of processing two JSON files in the <path>/chapter7/data/mobile directory and specify a watermark of 10 minutes. Listing 7-8 shows the data in those two files. The data is set up so that each row in the file1.json file falls into its own 10-minute window. The first row in the file2.json file falls into the 10:20:00 to 10:30:00 window, and even though it arrives late, its timestamp still falls within an acceptable threshold, so it is processed.

The last row in the file2.json file is a simulation of late data where its timestamp is in the 10:10:00 to 10:20:00 window, and since that falls outside the watermark threshold, it is ignored and not processed.

Listing 7-8. Mobile Event Data in Two JSON Files

```
// file1.json
{"id":"phone1","action":"open","ts":"2018-03-02T10:15:33"}
{"id":"phone2","action":"open","ts":"2018-03-02T10:22:35"}
{"id":"phone3","action":"open","ts":"2018-03-02T10:33:50"}

// file2.json
{"id":"phone4","action":"open","ts":"2018-03-02T10:29:35"}
{"id":"phone5","action":"open","ts":"2018-03-02T10:11:35"}
```

To simulate the processing, first create a directory called input under the <path>/ chapter7/data directory. Then run the code in Listing 7-9. The next step is to copy the file1.json file to the input directory and examine the output. The final step is to copy the file2.json file to the input directory and examine the output.

Listing 7-9. Code for Processing Mobile Data Events with Late Arrival

```
import org.apache.spark.sql.types._
import org.apache.spark.sql.functions._

val mobileSchema = new StructType().add("id", StringType, false)
                                   .add("action", StringType, false)
                                   .add("ts", TimestampType, false)

val mobileSSDF = spark.readStream.schema(mobileSchema)
                    .json("<path>/book/chapter7/data/input")

// setup a streaming DataFrame with a watermark and group by ts and action
column.
val windowCountDF = mobileSSDF.withWatermark("ts", "10 minutes")
                            .groupBy(window($"ts",
                                    "10 minutes"), $"action")
                                .count
```

```
val mobileMemorySQ = windowCountDF.writeStream
                                  .format("console")
                                  .option("truncate", "false")
                                  .outputMode("update")
                                  .start()

// the output from processing file1.json
// as expected each row falls into its own window
+---------------------------------------------+--------+------+
|                    window                   | action| count|
+---------------------------------------------+--------+------+
|    [2018-03-02 10:20:00, 2018-03-02 10:30:00]|  open  |  1   |
|    [2018-03-02 10:30:00, 2018-03-02 10:40:00]|  open  |  1   |
|    [2018-03-02 10:10:00, 2018-03-02 10:20:00]|  open  |  1   |
+---------------------------------------------+--------+------+

// the output from processing file2.json
// notice the count for window 10:20 to 10:30 is now updated to 2
// and there was no change to the window 10:10:00 and 10:20:00
+---------------------------------------------+---------+------+
|                    window                   | action  | count|
+---------------------------------------------+---------+------+
|    [2018-03-02 10:20:00, 2018-03-02 10:30:00]|  open   |  2   |
+---------------------------------------------+---------+------+
```

Since the timestamp of the last line in the file2.json file falls outside the 10-minute watermark threshold, it was not processed. If the call to watermark API is removed, the output looks something like Listing 7-10. The count of windows 10:10 and 10:20 is updated to 2.

Listing 7-10. Output of Removing the Call to Watermark API

```
+---------------------------------------------+--------+------+
|                    window                   | action| count|
+---------------------------------------------+--------+------+
|    [2018-03-02 10:20:00, 2018-03-02 10:30:00]|  open  |     2|
|    [2018-03-02 10:10:00, 2018-03-02 10:20:00]|  open  |     2|
+---------------------------------------------+--------+------+
```

Watermark is a useful feature, so it is important to understand the conditions under which the aggregation state is properly cleaned up.

- The output mode can't be the complete mode and must be in either update or append mode. The reason is the semantics of the complete mode dictate that all aggregate data must be maintained, and to not violate those semantics; the watermark can't drop any intermediate state.

- The aggregation via the groupBy transformation must be directly on the event time column or a window on the event time column.

- The event time column specified in the Watermark API and the groupBy transformation must be the same one.

- When setting up a streaming DataFrame, the Watermark API must be called before the groupBy transformation; otherwise, it is ignored.

Arbitrary Stateful Processing

The intermediate state of aggregations by key or event window is automatically maintained by Structured Streaming. However, not all event-time-based processing can be satisfied by simply aggregating on one or more columns and with or without windowing. For example, you want to send out an alert, email, or pager when three consecutive temperature readings with a value greater than 100 degrees are seen in the IoT temperature dataset.

Maintaining user sessions is anoher example, where the session length is not determined by a fixed amount of time but rather by the user's activities and the lack thereof. To solve these two examples and similar use cases, you need to apply arbitrary processing logic on each group of data, control the window length for each group of data, and maintain an arbitrary state across trigger points. This is where Structured Streaming arbitrary state processing comes in.

Arbitrary Stateful Processing with Structured Streaming

To enable flexible and arbitrary stateful processing, Structure Streaming provides a callback mechanism. It takes care of ensuring the intermediate state is maintained and stored in a fault-tolerant manner. The callback mechanism enables you to provide a user-defined function with your custom state management logic, and Structured Streaming calls it at the appropriate time This style of processing essentially boils down to the ability to perform one of the two following tasks, which are illustrated in Figure 7-4.

- Map over groups of data, apply arbitrary processing on each group, and generate only a single row per group.

- Map over groups of data, apply arbitrary processing on each group, and generate any number of rows per group, including none.

Structure Streaming provides a specific API for each of these tasks. For the first task, the API is called `mapGroupsWithState,` and for the second one, the API is called `flatMapGroupsWithState`. These APIs are available starting with Spark 2.2 and only in Scala and Java.

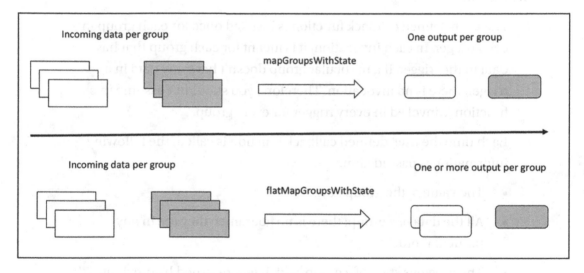

Figure 7-4. *Visual description of the two arbitrary stateful processing tasks*

When working a callback mechanism, it is important to clearly understand the contract between the framework and callback function regarding the input arguments, when and how often it gets called. In this case, the sequence goes as follows.

- To perform arbitrary stateful processing on a streaming DataFrame, you must first specify the grouping by calling the `groupByKey` transformation and provide a group by column; it then returns an instance of `KeyValueGroupedDataset` class.

- From an instance of KeyValueGroupedDataset class, you can call either mapGroupsWithState or flatMapGroupsWithState function. Each API requires a different set of input parameters.

- When calling mapGroupsWithState function, you need to provide the timeout type and a user-defined callback function. The timeout part is explained in a moment.

- When calling flatMapGroupsWithState function, you need to provide an output mode, the timeout type, and a user-defined callback function. Both the output mode and timeout parts are explained in a moment.

The following is the contract between Structured Streaming and the user-defined callback function.

- The user-defined callback function is invoked once for each group in each trigger. In each invocation, it is meant for each group that has data in the trigger. If a particular group doesn't have any data in a trigger, there is no invocation. Therefore, you shouldn't assume this function is invoked in every trigger for every group.

- Each time the user-defined callback function is called, the following information is passed along.

  - The value of the group key.

  - All the data of a group; there is no guarantee they are in any particular order.

  - The previous state of a group, which was returned by previous invocation of the same group. A group state is managed by a state holder class called GroupState. When there is a need to update the state of a group, you must call the update function of this class with the new state. A user-defined class defines the information in the state for each group. When calling the update function, the provided user-defined state can't be null.

As you learned in Chapter 6, only certain output modes are allowed whenever there is a need to maintain the intermediate state. As of Spark 2.3, only the update output mode is supported when calling API `mapGroupsWithState` API; however, both append and update modes are supported when calling `flatMapGroupsWithState` API.

Handling State Timeouts

In the case of event-time aggregation with watermark, the timeout of the intermediate state is internally managed by Structured Streaming, and there isn't any way to influence it. On the other hand, Structured Streaming's arbitrary stateful processing provides the flexibility to control the intermediate state timeout. Since you can maintain an arbitrary state, it makes sense for the application logic to control the intermediate state timeout to meet specific use cases.

Structure Streaming stateful processing provides three different timeout types. The first one is based on processing time, and the second one is based on event time. The timeout type is configured at the global level, meaning it is for all the groups within a particular streaming DataFrame. The timeout amount can be configured for each individual group and can be changed at will. If the intermediate state is configured with a timeout, it is important to check whether it timed out before processing the given list of values in the callback function. A timeout is not needed in some use cases, and the third timeout type is designed for this scenario. The timeout type is defined in the `GroupStateTimeout` class. You specify the type when calling `mapGroupsWithState` or `flatMapGroupsWithState` function. The timeout duration is specified using either the `GroupState.setTimeoutDuration` or `GroupState.setTimeoutTimeStamp` function for processing time out and event time out, respectively.

Keen readers may be wondering what happens when an intermediate state of a specific group has timed out. The contract Structure Streaming provides regarding this situation is that it calls your user-defined callback function with an empty list of values and sets the flag `GroupState.hasTimedOut` to true.

Of the three timeout types, the event-time timeout is the most complicated one and is covered first. Event-time timeout implies that it is based on the time in the event and therefore setting a watermark in the streaming DataFrame via `DataFrame.withWatermark` is required. To control the time out per group, you need to provide a timestamp value to the `GroupState.setTimeoutTimestamp` function during the processing of a particular group. The intermediate state of a group is timed out when the watermark advances beyond the provided timestamp. In the user sessionization use case, as a user interacts with your website, the session is extended by simply updating the timeout timestamp based on the user's latest interaction time plus some threshold. This ensures that as long as a user interacts with your website, the user session remains active, and the intermediate data is not timed out.

The processing timeout type works in a similar fashion to the event-time timeout type; however, the difference is that it is based on the wall clock of the server, which is constantly advancing forward. To control the time out per group, you provide a time duration to the `GroupState.setTimeoutDuration` function during the processing of a particular group. The time duration can be something like one minute, one hour, or two days. The intermediate state of a group is timed out when the clock has advanced past the provided duration. Since this timeout type depends on the system clock, it is important to consider the scenario when the time zone changes or there is a clock skew.

This may be obvious to keen readers, but it is important to recognize that the user-defined callback function is not called when there is no incoming data in the stream for a while. In addition, the watermark does not advance, and the timeout function call does not happen.

At this point, you should have a good understanding of how arbitrary state processing in Structured Streaming works and which APIs are involved. The following section work through a couple of examples to demonstrate how to implement arbitrary state processing.

Arbitrary State Processing in Action

This section demonstrates the arbitrary state processing in Structured Streaming by working through two use cases.

- The first one is about extracting patterns out of the data center computer rack temperature data and maintaining the status of each rack in the intermediate state. Whenever three consecutive temperatures of 100 degrees or more are encountered, the rack status is upgraded to the warning level. This example uses the `mapGroupsWithState` API.

- The second example is user sessionization, which keeps track of the user state based on their interactions with a website. This example uses the `flatMapGroupsWithState` API.

Regardless of which API performs arbitrary state processing, a common set of setup steps is needed.

1. Define a few classes to represent the input data, the intermediate state, and the output.

2. Define two functions. The first one is the callback function for Structured Streaming to call. The second function contains the arbitrary state processing logic on the data of each group as well as the logic to maintain the state.

3. Decide on a timeout type and an appropriate value for it.

Extracting Patterns with mapGroupsWithState

This use case aims to identify a particular pattern in the data center computer rack temperature data. The pattern of interest is the three consecutive temperature readings of 100 degrees or more from the same rack. The time difference between two consecutive high-temperature readings must be within 60 seconds. When such a pattern is detected, the status of that rack is upgraded to a warning status. If the next incoming temperature reading falls below the 100-degree threshold, the rack status is downgraded to normal.

The data for this example is in the `<path>/chapter7/data/iot_pattern` directory, which consists of three files, and their content is shown in Listing 7-11. The content of `file1.json` shows the temperature of `rack1` is alternating between above and below 100 degrees. The `file2.json` file shows the temperature of rack2 is heating up. In the `file3.json` file, rack3 is heating up, but the temperature readings are more than one minute apart.

Listing 7-11. Temperature Data in file1.json, file2.json and file3.json

```
// file1.json
{"rack":"rack1","temperature":99.5,"ts":"2017-06-02T08:01:01"}
{"rack":"rack1","temperature":100.5,"ts":"2017-06-02T08:02:02"}
{"rack":"rack1","temperature":98.3,"ts":"2017-06-02T08:02:29"}
{"rack":"rack1","temperature":102.0,"ts":"2017-06-02T08:02:44"}

// file2.json
{"rack":"rack1","temperature":97.5,"ts":"2017-06-02T08:02:59"}
{"rack":"rack2","temperature":99.5,"ts":"2017-06-02T08:03:02"}
```

```
{"rack":"rack2","temperature":105.5,"ts":"2017-06-02T08:03:44"}
{"rack":"rack2","temperature":104.0,"ts":"2017-06-02T08:04:06"}
{"rack":"rack2","temperature":108.0,"ts":"2017-06-02T08:04:49"}

// file3.json
{"rack":"rack2","temperature":108.0,"ts":"2017-06-02T08:06:40"}
{"rack":"rack3","temperature":100.5,"ts":"2017-06-02T08:06:20"}
{"rack":"rack3","temperature":103.7,"ts":"2017-06-02T08:07:35"}
{"rack":"rack3","temperature":105.3,"ts":"2017-06-02T08:08:53"}
```

Next you prepare a few classes and two functions to apply the pattern detection logic to the previous data. For this use case, the rack temperature data input data is represented by the RackInfo class and both the intermediate state and output are represented by the same class called RackState. Listing 7-12 shows the code.

Listing 7-12. Scala Case Classes for the Input and Intermediate State

```
case class RackInfo(rack:String, temperature:Double,
                    ts:java.sql.Timestamp)

// notice the constructor arguments are defined to be modifiable so you can
   update them
// the lastTS variable is used to compare the time between previous and
current temperature reading
case class RackState(var rackId:String, var highTempCount:Int,
                     var status:String,
                     var lastTS:java.sql.Timestamp)
```

Next, you define two functions. The first one is called updateRackState, which contains the core logic of the event pattern detection about the three consecutive temperature readings within a certain amount of time. The second function is updateAcrossAllRackStatus, which is the callback function that is passed into the mapGroupsWithState API. It makes sure the rack temperature readings are processed according to the order of their event time. Listing 7-13 is the code.

Listing 7-13. the Functions for Performing Pattern Detection

```scala
import org.apache.spark.sql.streaming.GroupState

// contains the main logic to detect the temperature pattern described above

def updateRackState(rackState:RackState, rackInfo:RackInfo) : RackState = {
  // setup the conditions to decide whether to update the rack state
  val lastTS = Option(rackState.lastTS).getOrElse(rackInfo.ts)
  val withinTimeThreshold = (rackInfo.ts.getTime -
                              lastTS.getTime) <= 60000
  val meetCondition = if (rackState.highTempCount < 1) true
                      else withinTimeThreshold
  val greaterThanEqualTo100 = rackInfo.temperature >= 100.0

  (greaterThanEqualTo100, meetCondition) match {
    case (true, true) => {
      rackState.highTempCount = rackState.highTempCount + 1
      rackState.status = if (rackState.highTempCount >= 3)
                          "Warning" else "Normal"
    }
    case _ => {
      rackState.highTempCount = 0
      rackState.status = "Normal"
    }
  }
  rackState.lastTS = rackInfo.ts
  rackState
}

// call-back function to provide mapGroupsWithState API
def updateAcrossAllRackStatus(rackId:String,
                                inputs:Iterator[RackInfo],
                      oldState: GroupState[RackState]) : RackState = {

  // initialize rackState with previous state if exists, otherwise create
    a new state
  var rackState = if (oldState.exists) oldState.get
                  else RackState(rackId, 5, "", null)
```

```
// sort the inputs by timestamp in ascending order
inputs.toList.sortBy(_.ts.getTime).foreach( input => {
  rackState = updateRackState(rackState, input)
// very important to update the rackState in the state holder class
  GroupState
  oldState.update(rackState)
})
rackState
}
```

The setup step is now complete, and now you wire the callback function into the mapGroupsWithState in the Structured Streaming application in Listing 7-14. The steps to simulate the streaming data are similar to previous examples, as shown here.

1. Create a directory called input under the <path>/chapter7/data directory. Remove all files in this directory if it already exists.

2. Run the code in Listing 7-14.

3. Copy file1.json to the input directory, then observe the output. Repeat this same step with file2.json and file3.json.

Listing 7-14. Using Arbitrary State Processing to Detect Patterns in a Streaming Application

```
import org.apache.spark.sql.streaming.{GroupStateTimeout, OutputMode}
import org.apache.spark.sql.types._
import org.apache.spark.sql.functions._

// schema for the IoT data
val iotDataSchema = new StructType()
                      .add("rack",StringType, false)
                      .add("temperature", DoubleType, false)
                      .add("ts", TimestampType, false)

val iotSSDF = spark.readStream.schema(iotDataSchema)
                .json("<path>/chapter7/data/input")

val iotPatDF = iotSSDF.as[RackInfo].groupByKey(_.rack)
                    .mapGroupsWithState[RackState,RackState]
        (GroupStateTimeout.NoTimeout)(updateAcrossAllRackStatus)
```

```
// setup the output and start the streaming query
val iotPatternSQ = iotPatDF.writeStream.format("console")
                          .outputMode("update")
                          .start()
```

```
// after file3.json is copied over to "input" directory, run the line below
   stop the streaming query
iotPatternSQ.stop
```

```
// the output after processing file1.json
+--------+-------------------+---------+-------------------+
|  rackId|       highTempCount|   status|             lastTS|
+--------+-------------------+---------+-------------------+
|   rack1|                  1|   Normal| 2017-06-02 08:02:44|
+--------+-------------------+---------+-------------------+
```

```
// the output after processing file2.json
+--------+-------------------+---------+-------------------+
|  rackId|       highTempCount|   status|             lastTS|
+--------+-------------------+---------+-------------------+
|   rack1|                  0|   Normal| 2017-06-02 08:02:59|
|   rack2|                  3|  Warning| 2017-06-02 08:04:49|
+--------+-------------------+---------+-------------------+
```

```
// the output after processing file3.json
+--------+-------------------+---------+-------------------+
|  rackId|       highTempCount|   status|             lastTS|
+--------+-------------------+---------+-------------------+
|   rack3|                  1|   Normal| 2017-06-02 08:08:53|
|   rack2|                  0|   Normal| 2017-06-02 08:06:40|
+--------+-------------------+---------+-------------------+
```

rack1 has a few temperature readings with over 100 degrees; however, they are not consecutive, and therefore the output status is at a normal level. In the file2.json file, rack2 has three consecutive temperature readings with over 100 degrees, and the time gap between each one and the one before is less than 60 seconds; therefore, the status of rack2 is at a warning level. rack3 has three consecutive temperature readings with over 100 degrees; however, the time gap between each one and the one before is more than 60 seconds; therefore, its status is at a normal level.

User Sessionization with flatMapGroupsWithState

This use case performs user sessionization using `flatMapGroupsWithState` API, which supports the ability to output more than one row per group. In this example, the sessionization processing logic is based on user activities. A session is created when the `login` action is taken. A session ends when the `logout` action is taken. A session is automatically ended when there are no user activities for 30 minutes. You leverage the timeout feature to perform this detection. Whenever a session starts or ends, that information is sent to the output. The output information consists of user id, session start and end times, and the number of visited pages.

The data for this use case is in the `<path>/chapter7/data/sessionization` directory, and it has three files. Their content is shown in Listing 7-15. The `file1.json` file contains the activities of user1, and it includes a `login` action, but there is no `logout` action. The `file2.json` file contains all the activities of user2, including both `login` and `logout` actions. The `file3.json` file contains only the `login` action for user3. The timestamp of the user activities in three files is set up so that the user1 session times out when `file3.json` is processed. By then, the amount of time user1 has been idled is more than 30 minutes.

Listing 7-15. User Activity Data

```
// file1.json
{"user":"user1","action":"login","page":"page1", "ts":"2017-09-06T08:08:53"}
{"user":"user1","action":"click","page":"page2", "ts":"2017-09-06T08:10:11"}
{"user":"user1","action":"send","page":"page3", "ts":"2017-09-06T08:11:10"}

// file2.json
{"user":"user2","action":"login", "page":"page1", "ts":"2017-09-06T08:44:12"}
{"user":"user2","action":"view", "page":"page7", "ts":"2017-09-06T08:45:33"}
{"user":"user2","action":"view", "page":"page8", "ts":"2017-09-06T08:55:58"}
{"user":"user2","action":"view", "page":"page6", "ts":"2017-09-06T09:10:58"}
{"user":"user2","action":"logout","page":"page9", "ts":"2017-09-06T09:16:19"}

// file3.json
{"user":"user3","action":"login", "page":"page4", "ts":"2017-09-06T09:17:11"}
```

Next, you prepare a few classes and two functions to apply the user sessionization logic to the previous data. For this use case, the user activity input data is represented by the UserActivity class. The intermediate state of user session data is represented by the UserSessionState class, and the UserSessionInfo class represents the user session output. The code for all these three classes is shown in Listing 7-16.

Listing 7-16. Scala Case Classes for Input, Intermediate State, and Output

```scala
case class UserActivity(user:String, action:String,
                        page:String, ts:java.sql.Timestamp)
// the lastTS field is for storing the largest user activity timestamp and
   this information is used
// when setting the timeout value for each user session
case class UserSessionState(var user:String, var status:String,
                            var startTS:java.sql.Timestamp,
                            var endTS:java.sql.Timestamp,
                            var lastTS:java.sql.Timestamp,
                            var numPage:Int)
// the end time stamp is filled when the session has ended.
case class UserSessionInfo(userId:String, start:java.sql.Timestamp,
end:java.sql.Timestamp,  numPage:Int)
```

Next, you define two functions. The first one is called updateUserActivity, which is responsible for updating the user session state based on user activity. It also updates the session start or end time based on user action and the latest activity timestamp. The second function is called updateAcrossAllUserActivities. It is the callback function that is passed into the flatMapGroupsWithState function. This function has two main responsibilities. The first one is to handle the timeout of the intermediate session state, and it updates the user session end time when such a condition arises. The other responsibility is to determine when and what to send to the output. The desired output is to emit one row when a user session is started and another one when a user session is ended. Listing 7-17 is the logic of these two functions.

Listing 7-17. the Functions for Performing User Sessionization

```scala
import org.apache.spark.sql.streaming.GroupState
import scala.collection.mutable.ListBuffer

def updateUserActivity(userSessionState:UserSessionState,
userActivity:UserActivity) : UserSessionState = {
    userActivity.action match {
      case "login" => {
        userSessionState.startTS = userActivity.ts
        userSessionState.status = "Online"
      }
      case "logout" => {
        userSessionState.endTS = userActivity.ts
        userSessionState.status = "Offline"
      }
      case _ => {
        userSessionState.numPage += 1
        userSessionState.status = "Active"
      }
    }

    userSessionState.lastTS = userActivity.ts
    userSessionState
}

def updateAcrossAllUserActivities(user:String,
                  inputs:Iterator[UserActivity],
                  oldState: GroupState[UserSessionState]) :
                  Iterator[UserSessionInfo] = {

  var userSessionState = if (oldState.exists) oldState.get
                      else UserSessionState(user, "",
    new java.sql.Timestamp(System.currentTimeMillis), null, null, 0)

  var output = ListBuffer[UserSessionInfo]()
```

```scala
inputs.toList.sortBy(_.ts.getTime).foreach( userActivity => {
  userSessionState = updateUserActivity(userSessionState,
                        userActivity)
                      oldState.update(userSessionState)

  if (userActivity.action == "login") {
    output += UserSessionInfo(user, userSessionState.startTS,
                        userSessionState.endTS, 0)

  }
})

val sessionTimedOut = oldState.hasTimedOut
val sessionEnded = !Option(userSessionState.endTS).isEmpty
val shouldOutput = sessionTimedOut || sessionEnded

shouldOutput match {
  case true => {
      if (sessionTimedOut) {
          userSessionState.endTS =
new java.sql.Timestamp(oldState.getCurrentWatermarkMs)
      }
      oldState.remove()
      output += UserSessionInfo(user, userSessionState.startTS,
                        userSessionState.endTS,
                        userSessionState.numPage)

  }
  case _ => {
    // extend sesion
    oldState.update(userSessionState)                oldState.setTimeoutTime
                                                     stamp(userSessionState.
                                                     lastTS.getTime,
"30 minutes")
  }
}

output.iterator
}
```

313

Once the setup step is completed, the next step is to wire the callback function into the `flatMapGroupsWithState` function in the Structured Streaming application, as shown in Listing 7-18. This example leverages the timeout feature, so it is required to set up a watermark and event-time timeout type. The following are the steps to simulate the streaming data.

1. Create a directory called `input` under the `<path>/chapter7/data` directory. Make sure to remove all existing files in this directory if it already exists.

2. Run the code shown in Listing 7-17.

3. Copy file1.json to the input directory, then observe the output. Repeat these steps for `file2.json` and `file3.json`.

Listing 7-18. Using Arbitrary State Processing to Perform User Sessionization in a Streaming Application

```
import org.apache.spark.sql.streaming.{GroupStateTimeout, OutputMode}
import org.apache.spark.sql.types._
import org.apache.spark.sql.functions._

val userActivitySchema = new StructType()
                              .add("user", StringType, false)
                              .add("action", StringType, false)
                              .add("page", StringType, false)
                              .add("ts", TimestampType, false)

val userActivityDF = spark.readStream.schema(userActivitySchema)
                      .json("<path>/chapter7/data/input")

// convert to DataSet of type UserActivity
val userActivityDS = userActivityDF.withWatermark("ts", "30 minutes").
as[UserActivity]

// specify the event-time timeout type and wire in the call-back function
val userSessionDS = userActivityDS.groupByKey(_.user)
    .flatMapGroupsWithState[UserSessionState,UserSessionInfo]
    (OutputMode.Append,GroupStateTimeout.EventTimeTimeout)
    (updateAcrossAllUserActivities)
```

```
// setup the output and start the streaming query
val userSessionSQ = userSessionDS.writeStream
                                  .format("console")
                                  .option("truncate",false)
                                  .outputMode("append")
                                  .start()

// only run this line of code below after done copying over file3.json
userSessionSQ.stop
```

```
// the output after processing file1.json
+---------+---------------------------+-----+------------+
|  userId|           start           | end |    numPage|
+---------+---------------------------+-----+------------+
|  user1 |        2017-09-06 08:08:53| null|      0     |
+---------+---------------------------+-----+------------+
```

```
// the output after processing file2.json
+---------+---------------------------+---------------------+--------+
|  userId|          start            |               end | numPage|
+---------+---------------------------+---------------------+--------+
|  user2 |      2017-09-06 08:44:12| null                |      0|
|  user2 |      2017-09-06 08:44:12| 2017-09-06 09:16:19|      3|
+---------+---------------------------+---------------------+--------+
```

```
// the output after processing file3.json

+---------+-------------------+---------------------+------------+
|  userId|      start        |              end   |    numPage|
+---------+-------------------+---------------------+------------+
|  user1 | 2017-09-06 08:08:53| 2017-09-06 08:46:19|          2|
|  user3 | 2017-09-06 09:17:11| null               |          0|
+---------+-------------------+---------------------+------------+
```

After processing the user activities in file1.json, there should be one row in the output. This is expected because whenever the updateAcrossAllUserActivities function sees a login action in the user activities, it adds an instance of the UserSessionInfo class to the ListBuffer output. There are two rows in the output after processing file2.json. One is for the login action, and the other one is for the logout action. Now, file3.json contains only one user activity for user3 with action login, but the output contains two rows. The row for user1 is the result of detecting the user1 session has timed out, which means the watermark has passed the timeout value of that particular session due to the lack of activities.

As demonstrated in the previous two use cases, the arbitrary stateful processing feature in Structured Streaming provides flexible and powerful ways to apply user-defined processing logic on each group with total control of what and when to send out to the output.

Handling Duplicate Data

Deduplicating data is a common need while processing data, and it is fairly easy to do this in batch processing. It is more challenging in stream processing because of the unbounded nature of streaming data. The data duplication in streaming data happens when the data producers send the same data multiple times to combat the unreliable network connection or transmission failures.

Luckily, Structured Streaming makes it easy for streaming applications to perform data duplication, and therefore these applications can guarantee exactly-once processing by dropping duplicate data as they arrive. The data duplication feature Structured streaming provides can work in conjunction with or without watermark. One key thing to remember, when performing data duplication without specifying the watermark, the state Structured Streaming needs to maintain grow infinitely throughout the lifetime of your streaming application, and this may lead to out-of-memory issues. With watermarking, the late data that is older than the watermark is automatically dropped to avoid duplicates.

The API to instruct Structured Streaming to perform data deduplication is simple. It has only one input: a list of column names to use to uniquely identify each row. The value of these columns performs duplicate detection, and Structured Streaming stores them as a state. The sample data that demonstrates the data deduplication feature has the same schema as the mobile event data. The count aggregation is based on the grouping of the id column. Both the id and ts columns are used as the user-defined keys for deduplication. The data for this example is in <path>/chapter7/data/deduplication. It contains two files: file1.json and file2.json. The content of these files is displayed in Listing 7-19.

Listing 7-19. Sample Data for the Data Duplication Example

```
// file1.json - each line is unique in term of id and ts columns
{"id":"phone1","action":"open","ts":"2018-03-02T10:15:33"}
{"id":"phone2","action":"open","ts":"2018-03-02T10:22:35"}
{"id":"phone3","action":"open","ts":"2018-03-02T10:23:50"}

// file2.json - the first two lines are duplicate of the first two lines in
    file1.json above
// the third line is unique
// the fourth line is unique, but it arrives late, therefore it will not be
    processed
{"id":"phone1","action":"open","ts":"2018-03-02T10:15:33"}
{"id":"phone2","action":"open","ts":"2018-03-02T10:22:35"}
{"id":"phone4","action":"open","ts":"2018-03-02T10:29:35"}
{"id":"phone5","action":"open","ts":"2018-03-02T10:01:35"}
```

To simulate the data deduplication, first you create a directory called input under the <path>/chapter7/data directory. Then you run the code in Listing 7-20. The next step is to copy the file1.json file to the input directory, and examine the output. The final step is to copy the file2.json file to the input directory and examine the output.

Listing 7-20. Deduplicating Data Using dropDuplicates API

```
import org.apache.spark.sql.types._
import org.apache.spark.sql.functions._

val mobileDataSchema = new StructType()
                        .add("id", StringType, false)
                        .add("action", StringType, false)
                        .add("ts", TimestampType, false)

// mobileDataSchema is defined in previous example
val mobileDupSSDF = spark.readStream.schema(mobileDataSchema)
                .json("<path>/chapter7/data/deduplication")

val windowCountDupDF = mobileDupSSDF.withWatermark("ts",
                                        "10 minutes")
                        .dropDuplicates("id", "ts")
                        .groupBy("id").count
```

317

```
val mobileMemoryDupSQ = windowCountDupDF.writeStream
                                    .format("console")
                                    .option("truncate", "false")
                                    .outputMode("update")
                                    .start()
// output after copying file1.json to input directory
+---------+--------+
|   id    |  count|
+---------+--------+
|  phone3| 1       |
|  phone1| 1       |
|  phone2| 1       |
+---------+--------+

// output after coping file2.json to input directory
+---------+--------+
|   id    |  count |
+---------+--------+
|  phone4| 1       |
+---------+--------+
```

As expected, after `file2.json` is copied to the input directory, only one line is displayed in the console. The first two lines are duplicates of the first two lines in file1. json, so they were filtered out. The last line has a timestamp of 10:10, which is considered late data since the timestamp is older than the 10-minute watermark threshold. Therefore, it was not processed and dropped.

Fault Tolerance

One of the most important considerations when developing streaming applications and deploying them to production is handling failure recovery. According to Murphy's law, anything that can go wrong will go wrong. Machines will fail, and software will have bugs.

Luckily, Structured Streaming provides a way to restart or recover your streaming application when there is a failure, and it continues where it left off. To take advantage of this recovery mechanism, you need to configure your streaming applications to use checkpointing and write-ahead logs by specifying a checkpoint location when setting up streaming queries. Ideally, the checkpoint location should be a directory on a reliable and fault-tolerant file system such as HDFS or Amazon S3. Structure Streaming periodically saves all the progress information, such as the offset details of the data being processed and the intermediate state values to the checkpoint location. Specifying a checkpoint location for a streaming query is very straightforward. You just need to add an option to your streaming query with the name checkpointLocation and the name of the directory as the value. Listing 7-21 is an example.

Listing 7-21. Add the checkpointLocation Option to a Streaming Query

```
val userSessionSQ = userSessionDS.writeStream.format("console")
                            .option("truncate",false)
        .option("checkpointLocation","/reliable/location")
        .outputMode("append")
        .start()
```

If you peek into the specified checkpoint location, you should see the following subdirectories: commits, metadata, offsets, sources, stats. The information in these directories is specific to a particular streaming query; hence, each must use a different checkpoint location.

Like most software applications, streaming applications evolve over time because of the need to improve the processing logic or performance or fix bugs. It is important to keep in mind how this might affect the information saved in the checkpoint location and to know what changes are considered safe to make. Broadly speaking, there are two categories of changes. One is the change to streaming application code, and the other is the change to Spark runtime.

Streaming Application Code Change

The information in the checkpoint location is designed to be somewhat resilient to the changes to streaming applications. There are a few kinds of changes that are considered incompatible changes. The first one is about changing the way the aggregation is done, such as changing the key column, adding more key columns, or removing one of the existing key columns. The second one is changing the class structure used to store the intermediate state, for example, when a field is removed, or the type of a field is changed from string to integer. When incompatible changes are detected during a restart, Structured Streaming lets you know via an exception. In this case, you must either use a new checkpoint location or remove the content in the previous checkpoint location.

Spark Runtime Change

The checkpoint format is designed to be forward compatible so that streaming applications can restart from an old checkpoint across Spark minor patch versions or minor versions (i.e., upgrading from Spark 2.2.0 to 2.2.1 or from Spark 2.2.x to 2.3.x). The only exception to the rule is when there are critical bug fixes. It is good to know that it is clearly documented in the release notes when Spark introduces incompatible changes.

If it is not possible to start a streaming application with an existing checkpoint location because of incompatibility issues, you need to use a new checkpoint location. You may also need to seed your applications with some information about the offset to read data from.

Streaming Query Metrics and Monitoring

Like other long-running applications such as online services, it is important to have some insights into your streaming applications regarding their progress, incoming data rate, or the amount of memory consumed by the intermediate state. Structured Streaming provides a few APIs to extract recent execution progress and an asynchronous way to monitor all streaming queries in a streaming application.

Streaming Query Metrics

The most basic useful information about a streaming query at any moment in time is its current status. You can retrieve and display this information in a human-readable format by calling the StreamingQuery.status function. The returned object is of type StreamingQueryStatus, and it can easily convert the status information into JSON format. Listing 7-22 shows an example of what the status information looks like.

Listing 7-22. Query Status Information in JSON Format

```
// use a streaming query from the example above
userSessionSQ.status
```

```
// output
res11: org.apache.spark.sql.streaming.StreamingQueryStatus =
{
  "message" : "Waiting for data to arrive",
  "isDataAvailable" : false,
  "isTriggerActive" : false
}
```

The status provides very basic information about what's going on in a streaming query. To get additional details from recent progress like the incoming data rate, processing rate, watermark, the offsets of the data source, and some information about the intermediate state, you can call the StreamingQuery.recentProgress function. This function returns an array of StreamingQueryProgress instances, which can convert the information into JSON format. By default, each streaming query is configured to retain 100 progress updates, and this number can be changed by updating the Spark configuration called spark.sql.streaming.numRecentProgressUpdates. To see the most recent streaming query progress, you can call the StreamingQuery.lastProgress function. Listing 7-23 shows a sample of a streaming query progress.

Listing 7-23. Streaming Query Progress Details

```
{
  "id" : "9ba6691d-7612-4906-b64d-9153544d81e9",
  "runId" : "c6d79bee-a691-4d2f-9be2-c93f3a88eb0c",
  "name" : null,
  "timestamp" : "2018-04-23T17:20:12.023Z",
  "batchId" : 0,
  "numInputRows" : 3,
  "inputRowsPerSecond" : 250.0,
  "processedRowsPerSecond" : 1.728110599078341,
  "durationMs" : {
    "addBatch" : 1548,
    "getBatch" : 8,
    "getOffset" : 36,
    "queryPlanning" : 110,
    "triggerExecution" : 1736,
    "walCommit" : 26
  },
```

```
  "eventTime" : {
    "avg" : "2017-09-06T15:10:04.666Z",
    "max" : "2017-09-06T15:11:10.000Z",
    "min" : "2017-09-06T15:08:53.000Z",
    "watermark" : "1970-01-01T00:00:00.000Z"
  },
  "stateOperators" : [ {
    "numRowsTotal" : 1,
    "numRowsUpdated" : 1,
    "memoryUsedBytes" : 16127
  } ],
  "sources" : [ {
    "description" : "FileStreamSource[file:<path>/chapter7/data/input]",
    "startOffset" : null,
    "endOffset" : {
      "logOffset" : 0
    },
    "numInputRows" : 3,
    "inputRowsPerSecond" : 250.0,
    "processedRowsPerSecond" : 1.728110599078341
  } ],
  "sink" : {
    "description" : "org.apache.spark.sql.execution.streaming.
    ConsoleSinkProvider@37dc4031"
  }
}
```

There are a few important key metrics in this streaming progress status to pay attention to. The input rate represents the amount of incoming data flowing into a streaming application from an input source. The processing rate tells you how fast your streaming application can process the incoming data. In an ideal state, the processing rate should be higher than the input rate, and if that is not the case, you need to consider scaling up the number of nodes in the Spark cluster. If a streaming application is maintaining state either implicitly through the groupBy transformation or explicitly through the arbitrary state processing APIs, it is important to pay attention to the metrics under the stateOperators section.

Spark UI provides a rich set of metrics at the job, stage, and task level. Each trigger in a streaming application is mapped to a job in Spark UI, where the query plan and task durations can be easily inspected.

Note   The streaming query status and progress details are available through an instance of a streaming query. While you streaming application is running in production, you don't have access to those streaming queries. What if you want to see that information from a remote host? One option is to embed a small HTTP server in your streaming application and expose a few simple URLs to retrieve that information.

Monitoring Streaming Queries via Callback

Structured Streaming provides a callback mechanism to asynchronously receive events and progress of the streaming queries in a streaming application. This is done by registering an implementation of the StreamingQueryListener interface. This interface defines several callback methods to receive status about your streaming query, such as when it started, when there is progress, and when it is terminated. An implementation of this interface has total control of what to do with the provided information. One example of the implementation is sending this information to a Kafka topic or some other publish-subscribe system for offline analysis or another streaming application to process. Listing 7-24 contains a very simple implementation of the StreamingQueryListener interface. It simply prints the information out to the console.

Listing 7-24. a Simple Implementation of StreamingQueryListener Interface

```
import org.apache.spark.sql.streaming.StreamingQueryListener
import org.apache.spark.sql.streaming.StreamingQueryListener.{
                  QueryStartedEvent, QueryProgressEvent,
                  QueryTerminatedEvent}

class ConsoleStreamingQueryListener extends StreamingQueryListener {
  override def onQueryStarted(event: QueryStartedEvent): Unit = {
    println(s"streaming query started: ${event.id} -
                            ${event.name} - ${event.runId}")
  }
```

```
override def onQueryProgress(event: QueryProgressEvent): Unit = {
    println(s"streaming query progress: ${event.progress}")
}

override def onQueryTerminated(event: QueryTerminatedEvent): Unit = {
    println(s"streaming query terminated: ${event.id} -
                                        ${event.runId}")
  }
}
```

Once you implement `StreamingQueryListener`, the next step is to register it with `StreamQueryManager`, which can handle multiple listeners. Listing 7-25 shows how to register and unregister a listener.

Listing 7-25. Register and Unregister an Instance of StreamingQueryListener with StreamQueryManager

```
Val listener = new ConsoleStreamingQueryListener

// to register
spark.streams.addListener(listener)

// to unregister
spark.streams.removeListener(listener)
```

One thing to remember is each listener receives the streaming query events of all the streaming queries in a streaming application. If there is a need to apply event processing logic to a specific streaming query, you can use the streaming query name to identify which one is of interest.

Monitoring Streaming Queries via Visualization UI

Spark 3.0 introduced a new and simple way to monitor all streaming queries via the Structured Streaming tab of the Spark UI, as shown in Figure 7-5. The visualization UI was designed to help Spark application developers troubleshoot their Structured Streaming applications during the development phase and gain insights into the real-time metrics. The UI displays two different kinds of statistics.

- • The summary information of each streaming query

- • The statistics on each streaming query, including input rate, process rate, input rows, and batch duration

Figure 7-5. *Structured Streaming tab in Spark UI*

Streaming Query Summary Information

A Structured Streaming application can have more than one streaming query, and whenever one of those is started, it is listed in the Structured Streaming tab of the Spark UI. The summary information is available for both active and completed streaming queries but in a separate section. Figure 7-6 shows the summary information of two streaming queries.

The summary information table contains the basic information for each streaming query, including query name, status, ID, run ID, start time, query duration, and aggregation statistics, like average input rate and average process rate. A streaming query can be in one of these statuses: RUNNING, FINISHED, and FAILED. The Error column contains useful information about the exception details of a failed query.

Streaming Query

▾ Active Streaming Queries (1)

Page: 1   1 Pages. Jump to 1   . Show 100   items in a page. Go

Name	Status	ID	Run ID	Start Time ▾	Duration	Avg Input /sec	Avg Process /sec	Latest Batch	
<no name>	RUNNING	2706daf3-cecc-4236-a959-aae99782796c	759ff7c6-86b3-473b-9d87-1889fbd21766	2021/09/18 15:44:40	4 minutes 3 seconds	9.62	0.01	1	

Page: 1   1 Pages. Jump to 1   . Show 100   items in a page. Go

▾ Completed Streaming Queries (1)

Page: 1   1 Pages. Jump to 1   . Show 100   items in a page. Go

Name	Status	ID	Run ID	Start Time ▾	Duration	Avg Input /sec	Avg Process /sec	Latest Batch	Error
<no name>	FINISHED	5a91c75a-99cb-494c-9ec8-50888c8a0f02	3857158f-a5e8-478a-98ee-542c021d50d5	2021/09/18 15:45:02	3 minutes 34 seconds	10.99	0.01	1	-

Page: 1   1 Pages. Jump to 1   . Show 100   items in a page. Go

Figure 7-6. *Structured Streaming tab with stream query summary information*

You can view the detailed statistics of a particular streaming query by clicking the link in the Run ID column.

Streaming Query Detailed Statistics Information

The Streaming Query Statistics page shows useful metrics to gain insights into your streaming application's performance and health and debug issues. Figure 7-7 shows the detailed statistics of a sample streaming query.

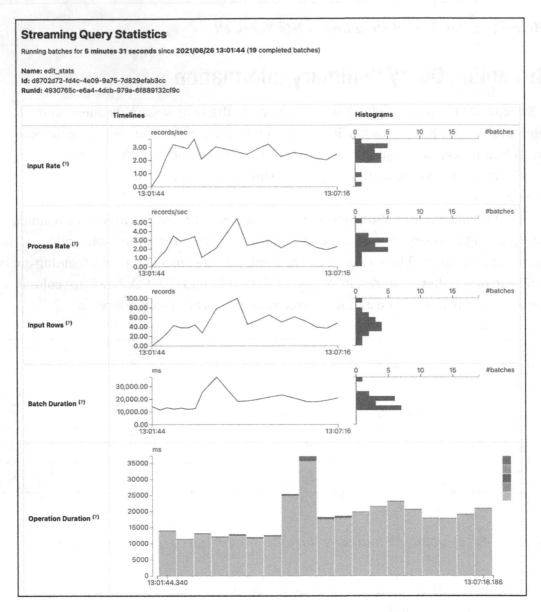

Figure 7-7. *Streaming query statistics example*

The following section describes the metrics shown in Figure 7-7. You can see a brief description of each metric by positioning your mouse over the question mark next to the metric name.

- **Input Rate**: The rate of data arriving across all the input sources of your streaming query. The rate is displayed as an aggregate, a single value to represent the rate of all the input sources.

- **Processing Rate**: The event processing rate at which the Structured Streaming engine is processing the incoming data. Like the preceding metric, it is an aggregate across all the input sources.

- **Batch Duration**: The processing time duration of each micro-batch

- **Operation Duration**: (in the context of a batch) The amount of time in milliseconds taken to perform the various operations.

  - addBatch: The time it takes to read, process, and write the batch's output to sink. This should take most of the time in the batch duration.

  - getBatch: The time it takes to prepare the logical query to read the input.

  - getOffset: The time it takes to query the input sources whether they have new input data.

  - walCommit: The time it takes to write the offsets to the metadata log.

  - queryPlanning: The time it takes to generate the execution plan.

Troubleshooting Streaming Query

With the availability of the detailed metrics of a streaming query, the next step is to leverage them to understand what's going on and that action to take to improve your Structured Streaming application performance. This section discusses two scenarios and shares a few suggestions to improve the streaming query performance.

The first scenario is when the input rate is much higher than the process rate metric. This is an indication that your streaming query is falling behind and unable to keep up with the data producers. The following are some possible actions to take.

- Add more execution resources such as increasing the number of executors

- Or increase the number of partitions to decrease the amount of work per partition

The second scenario is when the input rate is roughly the same as the process rate metric, but the batch duration metric is fairly high. This indicates that your streaming query is stable and able to the data producers, but the latency is high to process each batch. The following are some possible actions to take.

- Increase the parallelism of your streaming query

  - If the input source is Kafka, then increase the number of Kafka partitions

  - Increase the number of cores per Spark executor

The new Structured Streaming UI provides both summarized as well as detailed statistics about each streaming query. This is helpful for Spark developers to gain insights into their Structured Streaming application performance to take appropriate actions to address the performance issues.

Summary

The Spark Structured Streaming engine provides many advanced features and the flexibility to build complex and sophisticated streaming applications.

- Any serious streaming processing engine must support the ability to process incoming data by the event time. Structured Streaming not only supports the ability to do, but it also supports window aggregation based on fixed and sliding windows. In addition, it automatically maintains the intermediate state in a fault-tolerant manner.

- Maintaining the intermediate state introduces the risk of running out of memory as streaming applications process more and more data over a long time. Watermark was introduced to make it easier to reason about late data and remove no longer needed intermediate state.

- Arbitrary stateful processing enables a user-defined way of processing the values of each group and to maintain its intermediate state. Structured Streaming provides an easy way of doing this via callback API, and there is a flexibility in generating one or more rows per group to the output.

- Structured Streaming provides end-to-end exactly-once guarantee. This is achieved by using the checkpointing and the write ahead log mechanisms. Both of them can be turned on easily by providing a checkpoint location that resides on a fault-tolerant filesystem. Streaming applications can be easily restarted and pick up from where it left off before the failure by reading the information saved in the checkpoint location.

- Production streaming applications require the ability to have insights into the status and metrics of streaming queries. Structured Streaming provides a short summary of a streaming query status as well as the detailed metrics about incoming data rate, processing rate, and some details about the intermediate state memory consumption. To monitor the life cycle of streaming queries and their detailed progress, you can register one or more instances of StreamingQueryListener interface. The new Structure Streaming UI introduced in Spark 3.0 provides summarized and detailed statistics of each streaming query.

CHAPTER 8

Machine Learning with Spark

There has been a lot of excitement around artificial intelligence (AI), machine learning (ML), and deep learning (DL) in recent years. AI experts and researchers have predicted AI will radically transform the way humans live, work, and do business in the future. For businesses around the world, AI is one of the next steps in their journey of digital transformation, and some have made more progress than others in incorporating AI into their business strategies. Businesses expect AI to help solve their business problems efficiently and quickly and create new business value to increase their competitive advantages. Internet giants like Google, Amazon, Microsoft, Apple, and Facebook lead the pack in investing in, adopting, and incorporating AI into their product portfolio. In 2017, over $15 billion of venture capital (VC) money went into investing in AI-related start-up companies worldwide, and this trend is expected to continue.

AI is a broad area of computer science that attempts to make machines seem like they have intelligence. It is an audacious goal to help advance human mankind. One of the subfields within AI is machine learning, which focuses on teaching computers to learn without being explicitly programmed. The learning process involves extracting patterns from a large number of datasets using algorithms and building a model to explain the world. These algorithms can be categorized into different groups based on the task they are designed for. One trait these algorithms have in common is they learn through an iterative process of refining their internal parameters to achieve an optimal outcome.

Deep learning (DL) is one of the machine learning methods inspired by the way the human brain works, and it has proven to be good at learning complex patterns from data by representing them as a nested hierarchy of concepts. With the combination of the availability of large and curated datasets and the advancement in graphical processing units (GPUs), DL has proven to be effective at solving problems in areas such as object

© Hien Luu 2021
H. Luu, *Beginning Apache Spark 3*, https://doi.org/10.1007/978-1-4842-7383-8_8

recognition, image recognition, speech recognition, and machine translation. During an ImageNet image classification challenge, a computer system trained using the DL method beat a human at classifying images. The implication of this achievement and similar ones is that now computer systems can see, recognize objects, and hear at the same level as their creator. Figure 8-1 illustrates the relationship between AI, ML, and DL as well their timeline.

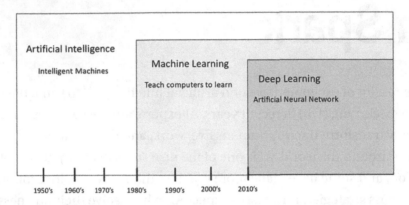

Figure 8-1. *Relationship between AI, ML, and DL and their timeline*

One of the motivations behind the creation of Spark is to help applications run iterative algorithms efficiently at scale. Over the last few versions of Spark, the MLlib library has steadily increased its offerings to make practical ML scalable and easy by providing a set of commonly used ML algorithms and a set of tools to facilitate the process of building and evaluating ML models.

To appreciate the features the MLlib library provides, it is necessary to have a fundamental understanding of the process of building ML applications. This chapter introduces the features and APIs available in the MLlib library.

Machine Learning Overview

This section provides a brief overview of machine learning and the ML application development process. It is not meant to be exhaustive; feel free to skip if you are already familiar with machine learning.

Machine learning is a vast and fascinating field of study, which combines parts of other fields of study, such as mathematics, statistics, and computer science. It teaches computers to learn patterns and derive insights from historical data, often for making decisions or predictions. Unlike traditional, hard-coded software, ML gives you only probabilistic outputs based on the imperfect data you provide. The more data you can provide to ML algorithms, the more accurate the output is. ML can solve much more interesting and difficult problems than traditional software can, and these problems are not specific to any industry or business domain. Examples of relevant areas are image recognition, speech recognition, language translation, fraud detection, product recommendation, robotics, autonomous driving car, speeding up the drug discovery process, medical diagnosis, customer churn prediction, recommendations, and many more.

Given that the goal of AI is to make machines seem like they have intelligence, one of the best ways to measure that is by comparing machine intelligence against human intelligence.

In recent decades, there are a few well-known and publicized demonstrations of such comparisons. The first was a computer system called Deep Blue that defeated the world chess champion in 1997 under strict tournament regulations. This example demonstrates that computer machines can think faster and better than humans in the game with a vast but limited set of possible moves.

The second one is about a computer system called Watson that competed on a *Jeopardy* game show against two legendary champions in 2011 and won the first prize of $1 million. This example demonstrates that computer machines can understand human language in a specific question-and-answer structure and then tap into its vast knowledge base to develop probabilistic answers.

The third one is about a computer program called AlphGo that defeated a world-champion Go player in a historic match in 2016. This example demonstrates a great leap in the advancement in the AI field. Go is a complex board game that requires intuition, creative and strategic thinking. It is not feasible to perform an exhaustive search move because the number of possible moves it has is greater than the number of atoms in the universe.

Machine Learning Terminologies

Before going deeper into ML, it is important to learn a few basic terminologies in this field. This is helpful in future sections when these terminologies are referenced. To make it easier to understand these terminologies, the explanations are provided in the canonical ML example called *spam email classification*.

- **Observation** is a term comes from the statistics field. An *observation* is an instance of the entity that is used for learning. For example, emails are considered observations.

- **Label** is a value to label an observation. For example, "spam" or "not spam" are two possible values used to label emails.

- **Features** are important attributes about observations that most likely have the strongest influence on the prediction output—for example, email sender IP address, the number words, and the number of capitalized words.

- **Training data** is a portion of the observations that train an ML algorithm to produce a model. A general practice is to split the collected data into three portions: training data, validation data, and test data. The test data portion is roughly about 70% or 80% of the original dataset.

- **Validation data** is a portion of the observations that evaluate the performance of the ML model during the model tuning process.

- **Test data** is a portion of the observations that evaluate the performance of the ML model after the tuning process is finalized.

- **ML algorithm** is a collection of steps that run iteratively to extract insights or patterns from given test data. The main goal of an ML algorithm is to learn a mapping from inputs to outputs. A well-known set of ML algorithms is available for you to choose from. The challenge is in selecting the right algorithm to solve a particular ML problem. For a spam email detection problem, you might pick the naïve Bayes algorithm.

- **Model**: After an ML algorithm learns from the given input data, it produces a model. You then use a model to perform predictions or make decisions on the new data. A model is represented by a mathematical formula. The goal is to produce a generalized model that performs well against any new data it has not seen before.

The relationship between an ML algorithm, data, and model is best illustrated in Figure 8-2.

$$model = algorithm(data)$$

Figure 8-2. *Relationship between ML algorithm, data, and model*

One important point to remember when applying machine learning is to never train an ML algorithm with the test data because that defeats the purpose of producing a generalized ML model. Another important point to note is ML is a vast field, and as you dig deeper into this field, you discover many more terminologies and concepts. Hopefully, this basic set of terminologies help you get started in this journey of learning ML.

Machine Learning Types

ML is about teaching machines to learn patterns from data for making decisions or predictions. These tasks are widely applicable to many different types of problems, where each problem type requires a different way of learning. There are three types of learning, which are depicted in Figure 8-3.

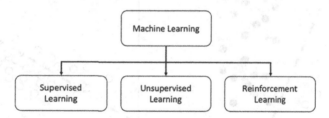

Figure 8-3. *Different machine learning types*

Supervised Learning

Among the three different learning types, this one is widely used and more popular because it can help solve a large class of problems in classification and regression.

Classification is about classifying the observations into one of the discrete or categorical classes of the label. Examples of classification problems include predicting whether an email is a spam email or not; whether a product review is positive or negative; whether an image contains a dog, cat, dolphin, or bird; whether the topic of a news article is about sports, medicine, politics, or religion; whether a particular handwritten digit is a 1 or 2; and whether the Q4 revenue met expectations. When the classification result has exactly two discrete values, that is called *binary classification*. When it has more than two discrete values, that is called *multiclass classification*.

Regression is about predicting real values from observations. Unlike classification, the predicted value is not discrete, but rather it is continuous. Examples of regression problems include predicting the house price based on their location and size, predicting the stock price of a company, predicting a person's income based on the background and education of a set of people, and so on.

One key distinguishing factor between this type of learning from the others is each observation in the training data must contain a label, whether that is discrete or continuous. In other words, the correct answers are provided to the algorithm to learn by iterating and incrementally improving its predictions on the training data. It stops once the acceptable error margin between the predicted value and actual value is achieved.

A simple mental model to distinguish classification from regression is that the former is about separating the data into various buckets, and the latter is about fitting the best line to the data. Figure 8-4 shows the visual representation of this mental model.

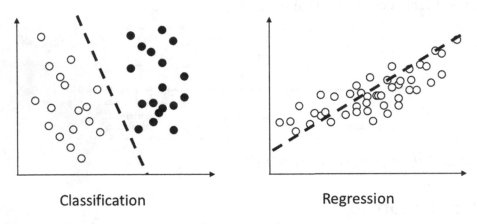

Classification   Regression

Figure 8-4. *Mental model of classification and regression*

A large collection of algorithms was designed to solve the classification and regression machine learning problems. This chapter touches on a few supported in the Spark MLlib component, as listed in Table 8-1.

Table 8-1. *Supervised Learning Algorithms in MLlib*

Tasks	Algorithms
Classification	Logistic Regression
	Decision Tree
	Random Forest
	Gradient-boosted Tree
	Linear Support Vector Machine
	Naïve Bayes
Regression	Linear Regression
	Generalized Linear Regression
	Decision Tree Regression
	Random Forest Regression
	Gradient-boosted Regression

Unsupervised Learning

The name of this learning method implies there is no supervision; in other words, the data trains the ML algorithm doesn't contain the labels. This learning type is designed to solve a different class of problems, such as discovering the hidden structure or patterns inside the data, and it is up to us, the human, to interpret the meaning behind those insights. One of the hidden structures, called *clustering*, is useful for deriving meaningful relationships or similarities between the observations within the clusters. Figure 8-5 depicts examples of clusters.

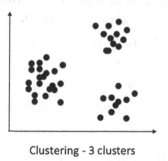

Clustering - 3 clusters

Figure 8-5. *Visualization of clustering*

As it turns out, there are many practical problems that this type of learning method can solve. Let's say there is a large collection of documents and no prior knowledge of which topic a particular document belongs to. You can use unsupervised learning to discover the clusters of related documents, and from there, you can assign a topic to each cluster. Another interesting and common problem that unsupervised learning can help solve is credit card fraud detection. After grouping user credit card transactions into clusters, it is not too difficult to spot the outliers, representing abnormal credit card transactions after a thief stole it. Table 8-2 lists the supported unsupervised learning algorithms in Spark.

Table 8-2. *Unsupervised Learning Algorithms in MLlib*

Tasks	Algorithms
Clustering	K-means
	Latent Dirichlet Allocation
	Bisecting *k*-means
	Gaussian

Reinforcement Learning

Unlike the first two types of learning, this one doesn't learn from data. Instead, it learns from interacting with an environment through a series of actions and the feedback it receives. Based on the feedback, it makes adjustments to move closer to its goal of maximizing some reward. In other words, it learns from its own experience.

Until recently, this type of learning hasn't gotten as much attention as the first two because it has not had significant practical success beyond computer games. In 2016, Google DeepMind was able to successfully apply this learning type to play an Atari game and then incorporate it into its AlphGo program, which defeated a world champion in the game of Go.

At this point, Spark MLlib doesn't include any reinforcement learning algorithms. The next sections focus on the first two types of learnings.

Note   The term *supervised* metaphorically refers to a teacher (human) who "supervises" the learner, which is the ML algorithm, by specifically providing the answers (labels) along with a set of examples (training data).

Machine Learning Development Process

To be effective at applying machine learning to develop intelligent applications, you should consider studying and adopting is a set of best practices that most ML practitioners follow. It has been said that effectively applying machine learning is a craft—half science and half art. Fortunately, a well-known and structured process consists of a series of steps to help provide reasonable repeatability and consistency, which is depicted in Figure 8-6.

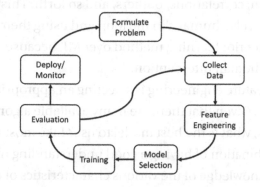

Figure 8-6. *Machine learning application development process*

The first step in this process is to clearly understand the business objective or challenge you think ML can help you with. It is beneficial to evaluate alternative solutions to ML to understand the cost and trade-offs. Sometimes it is faster to go with a simple rule-based solution to start with. If there is strong evidence that ML is a better choice to deliver valuable business insights efficiently, quickly, then you would proceed to the next step, which is to establish a set of success metrics that you and your stakeholders agree on.

Success metrics establish successful criteria on an ML project from a business perspective. They are measurable as well as directly related to business success. Examples of metrics are increasing customer conversion by a certain percentage, increasing the advertisement click-through rate by a certain amount, and increasing revenue by a certain amount of dollars. Success metrics are also helpful in deciding when to abandon an ML project due to cost or if it is not producing the expected gain.

After the success metrics are identified, the next step is to identify and collect the appropriate amount of data to train the ML algorithm. The quality and quantity of the collected data directly impacts the performance of the trained ML model. One important point to keep in mind is to make sure the collected data represents the problem you are trying to solve. The phrase "garbage in, garbage out" is still applicable in characterizing a key limitation in ML.

Featuring engineering is one of the most important and time-consuming steps in this process. It is mainly about data cleaning and using domain knowledge to identify key attributes or features in the observations to help the ML algorithms to learn the direct relationship between training data and provided labels. The data cleaning task is usually done using the exploratory data analysis framework to better understand the data in terms of data distribution, correlations, outliers, and so forth. This step is an expensive one due to the need to involve humans in the loop and using their domain knowledge. DL has shown to be a superior learning method over ML because it can automatically extract features without human intervention.

The next step after feature engineering is selecting an appropriate ML model or algorithm and training it. Given that there are many available algorithms to solve similar ML tasks, the question is, what is the best model to use? Like most things, deciding on the best one requires a combination of having a good understanding of the problem at hand, having a good working knowledge of the various characteristics of each algorithm, and having the experience to apply them to similar problems in the past. In other words, it is half science and half art when it comes to selecting the best algorithm. It requires some experimentation to arrive at the best algorithm. Once an algorithm is selected, let it learn from the features produced in the feature engineering step. The output of the training step is a model, which you then proceed to perform a model evaluation to see how well it performs. The goal of all the previous steps leading up to this one is to produce a generalized model, meaning how well it performs on data it has never seen before.

Another important step in the ML development process is the model evaluation task. It is both necessary and challenging. This step aims to not only answer the question of how well a model performs but also to know when to stop tuning the model because its performance has reached the established success metrics. The evaluation process can be done offline or online. The former case refers to evaluating the model using the training data, and the latter refers to evaluating the model using production or new data. There is a set of commonly used metrics to understand model performance: precision, recalls, F1 score, and AUC.

The art portion of this step is to understand which metrics are applicable for certain ML tasks. The model performance results determine whether to proceed to the production deployment step or to go back to the step of collecting more data or different types of data.

This information is meant to provide an overview of the ML development process and is not comprehensive. It can easily take a whole chapter to adequately cover the inner details of and best practices.

Spark Machine Learning Library

The remaining sections of this chapter cover the main features in the Spark MLlib component and provide examples of applying provided ML algorithms in Spark to each of the following ML tasks: classification, regression, clustering, and recommendation.

Note   In the Python world, scikit-learn is one of the most popular open source machine learning libraries. It is built on top of the NumPy, SciPy, and matplotlib libraries. It provides a set of supervised and unsupervised learning algorithms. It is designed to be a simple and efficient library, and it is a perfect one to learn and practice machine learning on a single machine. The moment the data size exceeds the storage capacity of a single machine, that's when it is time to switch to Spark MLlib.

There are many available ML libraries out there to choose from to train ML models in recent years. In the era of big data, there are two reasons to pick Spark MLlib over the other options. The first one is ease of use. Spark SQL provides a very user-friendly way of performing exploratory data analysis. The MLlib library provides a means to build, manage and persist complex ML pipelines. The second reason is about training ML at scale. The combination of Spark unified data analytic engine and the MLlib library can support training machine learning models with billions of observations and thousands of features.

Machine Learning Pipelines

The ML process is essentially a pipeline consisting of a series of steps that run sequentially. The pipeline usually needs to run multiple times to produce an optimal model. To make practical machine learning easy, Spark MLlib provides a set of abstractions to help simplify the steps of data cleaning, featuring engineering, training model, model tuning, and evaluation, and to organize them into a pipeline to make it easy to understand, maintain and repeat. The pipeline concept is inspired by the scikit-learn library.

There are four main abstractions to form an end-to-end ML pipeline: transformers, estimators, evaluators, and pipelines. They provide a set of standard interfaces to make it easy to work with and understand another data scientist's pipeline. Figure 8-7 depicts the similarity between the core steps in the ML process and the main abstractions MLlib provides.

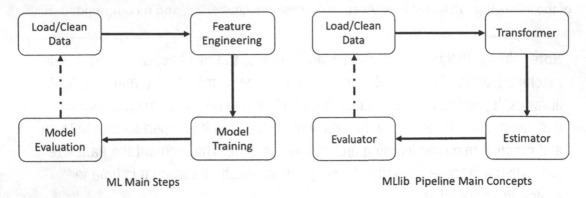

Figure 8-7. *Similarity between ML main steps and MLlib pipeline main concepts*

The one thing in common across these abstractions is that the type of input and output is mostly DataFrames, which means you need to convert the input data into a DataFrame to work with these abstractions.

Note   Like other components within the Spark unified data analytics engine, MLlib is switching to DataFrame-based APIs to provide more user-friendly APIs and take advantage of Spark SQL engine's optimizations. The new APIs are available in the org.apache.spark.ml package. The first MLlib version was developed on the RDD-based APIs, and it is still supported, but it is in maintenance mode only. The old APIs are available in the org.apache.spark.mllib package. Once the feature parity is reached, then the RDD-based APIs are deprecated.

Transformers

Transformers are designed to transform the data in the DataFrame by manipulating one or more columns during the feature engineering and model evaluation steps. The transforming process is in the context of building features that are consumed by the ML algorithm to learn. This process usually involves adding or removing columns (features), converting the column values from text to numerical values, or normalizing the values of a particular column.

There is a strict requirement about working with ML algorithms in MLlib; they require all features to be in Double data type, including the label.

From a technical perspective, a transformer has a `transform` function that performs transformations on the input columns, and the result is stored in the output column. The input column and output column names can be specified during the construction of a transformer. If they are not specified, the default column names are used. Figure 8-8 depicts what a transformer looks like; the shaded column in DF1 represents the input column. The darker shaded column in DF2 represents the output column.

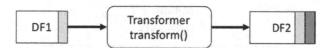

Figure 8-8. *Transformer input and output*

Each column data type needs a different set of data transformers. MLlib provides roughly about 30 transformers. Table 8-3 lists the various transformers for each type of data transformation.

Table 8-3. *Transformers for Different Transformation Types*

Type	Transformers
General	SQL Transformer
	VectorAssembler
Numeric Data	Bucketizer
	QuantileDiscretizer
	StandardScaler
	MixMaxScaler
	MaxAbsScaler
	Normalizer
Text Data	IndexToString
	OneHotEncoder
	Tokenizer, RegexTokenizer
	StopWordsRemover
	NGram
	HashingTF

This section discusses a few common transformers.

The `Binarizer` transformer simply transforms the values of one or more input columns into one or more output columns. The output value is either 0 and 1. The values that are less than or equal to the specified threshold are transformed to zero in the output column. For the values greater than the specified threshold, their value is transformed to one in the output column. The input column type must be double or VectorUDT. Listing 8-1 transforms the temperature column values into two buckets.

Listing 8-1. Use Binarizer Transformer Convert Temperature into Two Buckets

```
import org.apache.spark.ml.feature.Binarizer

val arrival_data = spark.createDataFrame(Seq(
            ("SFO", "B737", 18, 95.1, "late"),
            ("SEA", "A319", 5, 65.7, "ontime"),
            ("LAX", "B747", 15, 31.5, "late"),
```

```
                  ("ATL", "A319", 14, 40.5, "late") ))
              .toDF("origin", "model", "hour",
                    "temperature", "arrival")
val binarizer = new Binarizer().setInputCol("temperature")
                              .setOutputCol("freezing")
                              .setThreshold(35.6)

binarizer.transform(arrival_data).show

// show the current values of the parameters in binarizer transformer
binarizer.explainParams

inputCol: input column name (current: temperature)
outputCol: output column name (default: binarizer_60430bb4e97f__output,
current: freezing)
threshold: threshold used to binarize continuous features (default: 0.0,
current: 35.6)

// show the transformation result
binarizer.transform(arrival_data)
        .select("temperature", "freezing").show

+----------------+----------+
|     temperature|  freezing|
+----------------+----------+
|            95.1|       1.0|
|            65.7|       1.0|
|            31.5|       0.0|
|            40.5|       1.0|
+----------------+----------+
```

The Bucketizer transformer is a general version of the Binarizer where it can transform the column values into buckets of your choice. The way to control the number of buckets and the range of values of each bucket is by specifying a list of bucket borders in the form of an array of double values. This transformer is useful in the scenario where the values of a column are continuous, and you want to transform them into categorical values. For example, you have a column containing the income amount of each person who lives in a particular state, and you want to bucket their incomes into the following buckets: high income, middle income, and low income.

The value bucket border array must be of type double, and they must abide by the following requirements.

- The smallest bucket border value must be less than the minimum value in the input column in the DataFrame.

- The largest bucket border value must be greater than the maximum value in the input column in the DataFrame

- There must be at least three bucket borders in the input array, which creates two buckets.

In a person's income, it is easy to know the smallest income amount is 0. The smallest bucket border value can just be something less than 0. When it is not possible to predict the minimum column value, you can specify negative infinity. Similarly, when it is not possible to predict the maximum column value, then specify positive infinity.

Listing 8-2 is an example of using this transformer to bucket the temperature column into three buckets, which means the bucket border array must contain at least four values. It is sorted by the temperature column to make it easier to see.

Listing 8-2. Use Bucketizer Transformer Convert Temperature into Three Buckets

```
import org.apache.spark.ml.feature.Bucketizer

val bucketBorders = Array(-1.0, 32.0, 70.0, 150.0)
val bucketer = new Bucketizer().setSplits(bucketBorders)
                        .setInputCol("temperature")
                        .setOutputCol("intensity")

val output = bucketer.transform(arrival_data)
                .output.select("temperature", "intensity")
                .orderBy("temperature")
                .show
```

```
+----------------+-----------+
|     temperature|  intensity|
+----------------+-----------+
|            31.5|        0.0|
|            40.5|        1.0|
|            65.7|        1.0|
|            95.1|        2.0|
+----------------+-----------+
```

The OneHotEncoder transformer is commonly used when working with numeric categorical values. If the categorical values are of string type, you first apply the StringIndexer estimator and convert them to numerical type. OneHotEncoder essentially maps a numeric categorical value into a binary vector to purposely remove the implicit ranking of the numeric values. Listing 8-3 represents the student majors where each major is assigned an ordinal value, which suggests a certain major is higher than others. This transformer converts the ordinal value into a vector to remove such unintended bias during the ML training step. Listing 8-3 is an example of using this transformer.

Listing 8-3. Use OneHotEncoder Transformer the Ordinal Value of the Categorical Values

```
import org.apache.spark.ml.feature.OneHotEncoder

val student_major_data = spark.createDataFrame(
                         Seq(("John", "Math", 3),
                             ("Mary", "Engineering", 2),
                             ("Jeff", "Philosophy", 7),
                             ("Jane", "Math", 3),
                             ("Lyna", "Nursing", 4) ))
                         .toDF("user", "major", "majorIdx")

val oneHotEncoder = new OneHotEncoder().setInputCol("majorIdx")
                            .setOutputCol("majorVect")

oneHotEncoder.transform(student_major_data).show()
```

```
+------+---------------+-----------+---------------+
| user|          major|   majorIdx|      majorVect|
+------+---------------+-----------+---------------+
| John|           Math|         3|  (7,[3],[1.0])|
| Mary|    Engineering|         2|  (7,[2],[1.0])|
| Jeff|     Philosophy|         7|      (7,[],[])|
| Jane|           Math|         3|  (7,[3],[1.0])|
| Lyna|        Nursing|         4|  ( 7,[4],[1.0])|
+------+---------------+-----------+---------------+
```

Another common need when working with string categorical values is to convert them into ordinal values, which can be done using the StringIndexer estimator. This estimator is described in the "Estimator" section.

There are many interesting machine learning use cases where the input is in free-form text. It requires a few transformations to convert free-form text into numerical representation such that ML algorithms can consume it. Among them are tokenization and counting word frequency.

Most likely, you can guess what the Tokenizer transformer does. It performs the tokenization on a string of words separated by space and returns an array of words. If the delimiter is not space, then you can use RegexTokenizer with a specified delimiter. Listing 8-4 is an example of using the Tokenizer transformer.

Listing 8-4. Use Tokenizer Transformer to Perform Tokenization

```
import org.apache.spark.ml.feature.Tokenizer
import org.apache.spark.sql.functions._

val text_data = spark.createDataFrame(Seq(
            (1, "Spark is a unified data analytics engine"),
            (2, "It is fun to work with Spark"),
            (3, "There is a lot of exciting sessions at upcoming
                    Spark summit"),
            (4, "mllib transformer estimator evaluator
                    and pipelines"))).toDF("id", "line")

val tokenizer = new Tokenizer().setInputCol("line")
                               .setOutputCol("words")
```

```
val tokenized = tokenizer.transform(text_data)

tokenized.select("words")
        .withColumn("tokens", size(col("words")))
        .show(false)
```

```
+----------------------------------------------------------------+-------+
|                 words                                          | tokens|
+----------------------------------------------------------------+-------+
|[spark, is, a, unified, data, analytics, engine]                |      7|
|[spark, is cool, and, it, is, fun, to, work, with,              |     11|
|[there, is, a, lot, of, exciting, sessions, at, upcoming, spark, summit] |     11|
|[mllib, transformer, estimator, evaluator, and, pipelines]      |      6|
+----------------------------------------------------------------+-------+
```

Stop words are the commonly used words in a language. In the context of natural language processing or machine learning, stop words tend to add unnecessary noise and don't add any meaningful contributions. Therefore, they are usually removed immediately after the tokenization step. The StopWordsRemover transformer is designed to help with this effort.

As of Spark 2.3 version, the stop words for the following languages are included in Spark distribution: Danish, Dutch, English, Finnish, French, German, Hungarian, Italian, Norwegian, Portuguese, Russian, Spanish, Swedish, and Turkish. It is designed to be flexible so you can provide a set of stop words from a file.

To use the stop words in a particular language, you first call the StopWordsRemover. loadDefaultStopWords(<language in lower case>) to load them in and provide them to an instance of StopWordsRemover. Additionally, you can request this transformer to perform stop word filtering with case insensitive. Listing 8-5 is an example of using the StopWordsRemover transformer to remove English stop words.

Listing 8-5. Use StopWordsRemover Transformer to Remove English Stop Words

```
import org.apache.spark.ml.feature.StopWordsRemover

val enSWords = StopWordsRemover.loadDefaultStopWords("english")

val remover = new StopWordsRemover().setStopWords(enSWords)
                                    .setInputCol("words")
                                    .setOutputCol("filtered")

// use the tokenized from Listing 8-5 example
val cleanedTokens = remover.transform(tokenized)

cleanedTokens.select("words","filtered").show(false)
```

```
+---------------------------------------------------------------+----------------------------------------------------------+
|words                                                          |filtered                                                  |
+---------------------------------------------------------------+----------------------------------------------------------+
|[spark, is, a, unified, data, analytics, engine]              |[spark, unified, data, analytics, engine]                 |
|[spark, is, cool, and, it, is, fun, to, work, with, spark]    |[spark, cool, fun, work, spark]                           |
|[there, is, a, lot, of, exciting, sessions, at, upcoming, spark, summit]|[lot, exciting, sessions, upcoming, spark, summit]|
|[mllib, transformer, estimator, evaluator, and, pipelines]    |[mllib, transformer, estimator, evaluator, pipelines]     |
+---------------------------------------------------------------+----------------------------------------------------------+
```

The `HashingTF` transformer transforms a collection of words into numeric representations by computing the frequency of each word. Each word is mapped into an index by applying a hash function called MurmurHash 3. This approach is efficient, but it suffers from potential hash collisions, meaning multiple words may map into the same index. One way to minimize the collision is to specify a large number of buckets in the power of 2 to evenly distribute the words. Listing 8-6 feeds the filtered column from Listing 8-5 into the `HashingTF` transformer.

Listing 8-6. Use HashingTF Transformer to Transform Words into Numerical Representation Via Hashing and Counting

```
import org.apache.spark.ml.feature.HashingTF

val tf = new HashingTF().setInputCol("filtered")
                        .setOutputCol("TFOut")
                        .setNumFeatures(4096)
```

```
val tfResult = tf.transform(cleanedTokens)

tfResult.select("filtered", "TFOut").show(false)
```

```
+----------------------------------------------------+------------------------------------------------------------+
|filtered                                            |TFOut                                                       |
+----------------------------------------------------+------------------------------------------------------------+
|[spark, unified, data, analytics, engine]           |(4096,[991,1185,1461,3377,3717],[1.0,1.0,1.0,1.0,1.0])       |
|[spark, cool, fun, work, spark]                     |(4096,[251,1185,1575,2435],[1.0,2.0,1.0,1.0])               |
|[lot, exciting, sessions, upcoming, spark, summit]  |(4096,[724,1185,1255,1962,2966,3023],[1.0,1.0,1.0,1.0,1.0,1.0])|
|[mllib, transformer, estimator, evaluator, pipelines]|(4096,[994,2132,2697,3522,3894],[1.0,1.0,1.0,1.0,1.0])      |
+----------------------------------------------------+------------------------------------------------------------+
```

The last transformer this section covers is VectorAssembler, which combines a set of columns into a vector column. In machine learning terminology, that is the equivalent of combining individual features into a single-vector feature for ML algorithms to learn. The type of the individual input column must be one of the following types: numeric, boolean, or vector type. The output vector column contains the values of all the columns in the specified order. This transformer is used practically in every single ML pipeline, and its output is passed into an estimator. Listing 8-7 is an example of using a VectorAssembler transformer.

Listing 8-7. Use VectorAssembler Transformer to Combines Features into a Vector Feature

```
import org.apache.spark.ml.feature.VectorAssembler

val arrival_features  = spark.createDataFrame(Seq(
                                        (18, 95.1, true),
                                        (5, 65.7, true),
                                        (15, 31.5, false),
                                        (14, 40.5, false) ))
                      .toDF("hour", "temperature", "on_time")

val assembler = new VectorAssembler().setInputCols(
                  Array("hour", "temperature", "on_time"))
                              .setOutputCol("features")

val output = assembler.transform(arrival_features)
output.show
```

```
+-----+-----------------+-----------+------------------+
| hour|      temperature|    on_time|          features|
+-----+-----------------+-----------+------------------+
|   18|             95.1|       true|  [18.0,95.1,1.0]|
|    5|             65.7|       true|   [5.0,65.7,1.0]|
|   15|             31.5|      false|  [15.0,31.5,0.0]|
|   14|             40.5|      false|  [14.0,40.5,0.0]|
+-----+-----------------+-----------+------------------+
```

To make it easy to transform multiple columns at once, Spark version 3.0 added such support for these transformers: Binarizer, StringIndexer, and StopWordsRemover. Listing 8-8 shows a small example of transforming multiple columns with the Binarizer transformer. You have the option to specify a single threshold or multiple thresholds. If a single threshold is specified, then it is used for all the input columns. If multiple thresholds are specified, then the first threshold is used for the first input columns, and so on.

Listing 8-8. Transforming Multiple Columns With Binarizer Transformer

```
import org.apache.spark.ml.feature.Binarizer

val temp_data = spark.createDataFrame(
                    Seq((65.3,95.1),(60.7,99.1),
                        (75.3, 105.3)))
                    .toDF("morning_temp", "night_temp")

val temp_bin = new Binarizer()
        .setInputCols(Array("morning_temp", "night_temp"))
        .setOutputCols(Array("morning_oput","night_out"))
        .setThresholds(Array(65,96))

temp_bin.transform(temp_data).show
```

```
+------------+----------+------------+---------+
|morning_temp|night_temp|morning_oput|night_out|
+------------+----------+------------+---------+
|        65.3|      95.1|         1.0|      0.0|
|        60.7|      99.1|         0.0|      1.0|
|        75.3|     105.3|         1.0|      1.0|
+------------+----------+------------+---------+
```

Knowing how the transformers work and the available transformers in MLlib plays an important role in the feature engineering step of the ML development process. Generally, the output of a VectorAssembler transformer is consumed by an Estimator, which is covered in the next section.

Estimators

Estimators are an abstraction for an ML learning algorithm or any other algorithm that operates on data. It is rather confusing that an estimator can be one of two kinds of algorithms. An example of the first type is the ML algorithm called LinearRegression, which is used for a regression task to predict house prices. An example of the second algorithm kind is the StringIndexer, which encodes categorical values into indices. The index value for each categorical value is based on the frequency it appears in the entire input column of a DataFrame. At the high level, this kind of estimator transforms the values of a column into another column; however, it requires two passes over the entire DataFrame to produce the expected output.

From a technical perspective, an estimator has a fit function that applies an algorithm on the input column. The produced result is encapsulated in an object type called Model, which is a Transformer type. The input column and output column names can be specified during the construction of an estimator. Figure 8-9 depicts what an estimator looks like and its input and output.

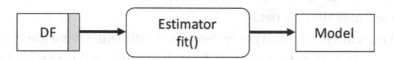

Figure 8-9. *Estimator and its input and output*

To give a sense of the two types of estimators, Table 8-4 provides a subset of the available estimators in MLlib.

Table 8-4. *Sample of Available Estimators in MLlib*

Type	Estimators
Machine Learning Algorithms	LogisticRegression
	DecisionTreeClassifier
	RandomForestClassifier
	LinearRegression
	RandomForestRegressor
	KMeans
	LDA
	BisectingKMeans
Data Transformation Algorithms	IDF
	RFormula
	StringIndexer
	OneHotEncoderEstimator
	StandardScaler
	MixMaxScaler
	MaxAbsScaler
	Word2Vec

The following section provides a few examples of commonly used estimators when working with text and numeric data.

RFormula is an interesting and general-purpose estimator where the transformation logic is expressed declaratively. It can handle both numeric and categorical values, and the output it produces is a vector of features. MLlib borrows the idea of this estimator from R language, and it supports only a subset of the operators available in R. The basic and supported operators are listed in Table 8-5. It takes time to understand the transformation language to take full advantage of the flexibility and power of the RFormula estimator.

Table 8-5. *Supported Operators in RFormula Transformer*

Operator	Description
~	Delimiter between the target and the terms
+	Concatenate terms
-	Remove a term
:	Interaction between other terms to create new features. Multiplication is used for numeric values and binarized for categorical values.
.	All columns except the target

Listing 8-9 specifies the label in the arrival column and the remaining columns as features. In addition, it creates a new feature using the interaction between the hour and temperature columns. Since these two columns are of numeric type, their values are multiplied.

Listing 8-9. Use RFomula Transformer to Create a Feature Vector

```
import org.apache.spark.ml.feature.RFormula

val arrival_data = spark.createDataFrame(Seq(
                    ("SFO", "B737", 18, 95.1, "late"),
                    ("SEA", "A319", 5, 65.7, "ontime"),
                    ("LAX", "B747", 15, 31.5, "late"),
                    ("ATL", "A319", 14, 40.5, "late") ))
                        .toDF("origin", "model", "hour",
                            "temperature", "arrival")

val formula = new RFormula().setFormula(
                    "arrival ~ . + hour:temperature")
                        .setFeaturesCol("features")
                        .setLabelCol("label")

// call fit function first, which returns a model (type of transformer),
then call transform
val output = formula.fit(arrival_data).transform(arrival_data)
```

```
output.select("*").show(false)
```

```
+------+-----+----+-----------+-------+----------------------------------------+-----+
|origin|model|hour|temperature|arrival|features                                |label|
+------+-----+----+-----------+-------+----------------------------------------+-----+
|SFO   |B737 |18  |95.1       |late   |(8,[0,5,6,7],[1.0,18.0,95.1,1711.8])    |0.0  |
|SEA   |A319 |5   |65.7       |ontime |[0.0,0.0,1.0,1.0,0.0,5.0,65.7,328.5]    |1.0  |
|LAX   |B747 |15  |31.5       |late   |(8,[4,5,6,7],[1.0,15.0,31.5,472.5])     |0.0  |
|ATL   |A319 |14  |40.5       |late   |[0.0,1.0,0.0,1.0,0.0,14.0,40.5,567.0]   |0.0  |
+------+-----+----+-----------+-------+----------------------------------------+-----+
```

One of the commonly used estimators when working with text is the IDF estimator. Its name is an acronym for *inverse document frequency*. This estimator is often used right after the text is tokenized and term frequency is computed. The idea behind this estimator is to compute the importance or weight of each word by counting the number of documents it appears in. The intuition behind this idea is that a word with a high occurrence and wide prevalence would be less important; for example, the word *the*. Inversely, a word with a high occurrence in only a few documents indicates higher importance; for example, the word *classification*. In the context of a DataFrame, a document is referring to a row. A keen reader would figure out that it requires going through every row to compute the importance of each word, and therefore IDF is an estimator, not a transformer. Listing 8-10 chains the Tokenizer and HashingTF transformers together with the IDF estimator. Unlike transformers, the estimators are eagerly evaluated, which means when the fit function is called, it triggers a Spark job.

Listing 8-10. Use IDF Estimator to Compute the Weight of Each Word

```
import org.apache.spark.ml.feature.Tokenizer
import org.apache.spark.ml.feature.HashingTF
import org.apache.spark.ml.feature.IDF

val text_data = spark.createDataFrame(Seq(
          (1, "Spark is a unified data analytics engine"),
          (2, "Spark is cool and it is fun to work with Spark"),
          (3, "There is a lot of exciting sessions at upcoming
              Spark summit"),
          (4, "mllib transformer estimator evaluator and
              pipelines")  )).toDF("id", "line")
```

```
val tokenizer = new Tokenizer().setInputCol("line")
                              .setOutputCol("words")

// the output column of the Tokenizer transformer is the input to HashingTF
val tf = new HashingTF().setInputCol("words")
                        .setOutputCol("wordFreqVect")
                        .setNumFeatures(4096)

val tfResult = tf.transform(tokenizer.transform(text_data))

// the output of the HashingTF transformer is the input to IDF estimator
val idf = new IDF().setInputCol("wordFreqVect")
                   .setOutputCol("features")

// since IDF is an estimator, call the fit function
val idfModel = idf.fit(tfResult)

// the returned object is a Model, which is of type Transformer
val weightedWords = idfModel.transform(tfResult)

weightedWords.select("label", "features").show(false)

weightedWords.printSchema
 |-- id: integer (nullable = false)
 |-- line: string (nullable = true)
 |-- words: array (nullable = true)
 |    |-- element: string (containsNull = true)
 |-- wordFreqVect: vector (nullable = true)
 |-- features: vector (nullable = true)

// the feature column contains a vector for the weight of each word, since
    it is long, the output is not included //below
weightedWords.select("wordFreqVect", "features").show(false)
```

When working with text data that contains categorical values, one commonly used estimator is the StringIndexer estimator. It encodes a categorical value into an index based on its frequencies such that the most frequent categorical value has an index value of 0 and so on. For this estimator to come up with an index value for a categorical value, it first must count the frequency of each categorical value and finally assign an index value to each one. To perform the counting and assign the index values, it must go through all the values of the input column from the beginning to the end of the DataFrame. If the input column is numeric, this estimator casts its string before computing its frequency.

Listing 8-11 provides an example of using the StringIndexer estimator to encode the movie genre.

Listing 8-11. StringIndex Estimator to Encode Movie Genre

```
import org.apache.spark.ml.feature.StringIndexer

val movie_data = spark.createDataFrame(Seq(
                                    (1, "Comedy"),
                                    (2, "Action"),
                                    (3, "Comedy"),
                                    (4, "Horror"),
                                    (5, "Action"),
                                    (6, "Comedy"))
                            ).toDF("id", "genre")

val movieIndexer = new StringIndexer().setInputCol("genre")
                                .setOutputCol("genreIdx")

// first fit the data
val movieIndexModel = movieIndexer.fit(movie_data)

// use returned transformer to transform the data
val indexedMovie = movieIndexModel.transform(movie_data)

indexedMovie.orderBy("genreIdx").show()

+---+-----------+------------+
| id|      genre|    genreIdx|
+---+-----------+------------+
|  3|     Comedy|         0.0|
|  6|     Comedy|         0.0|
|  1|     Comedy|         0.0|
|  5|     Action|         1.0|
```

```
|   2|     Action|          1.0|
|   4|     Horror|          2.0|
+---+----------+------------+
```

This estimator assigns the index based on the descending order of the frequency. This default behavior can be easily changed to ascending order of the frequency. It supports two other ordering types: descending alphabet and ascending alphabet. To change the default ordering type, you simply call the setStringOrderType("<ordering type>") function with one of the following values: frequencyDesc, frequencyAsc, alphabetDesc, and alphabetAsc.

With Spark version 3.0, the StringIndexer estimator can support encoding multiple columns of categorical value in a DataFrame. When there is such a need, you can simply call the setInputCols function to specify the input column names to encode and correspondingly specify the output column names by calling the setOutputCols function. Listing 8-12 provides an example of using the StringIndexer estimator to encode multiple columns.

Listing 8-12. StringIndex Estimator to Encode Multiple Columns

```
import org.apache.spark.ml.feature.StringIndexer

val movie_data2 = spark.createDataFrame(Seq(
                              (1, "Comedy", "G"),
                              (2, "Action", "PG"),
                              (3, "Comedy", "NC-17"),
                              (4, "Horror", "PG-13"))
                    ).toDF("id", "genre", "rating")

val movieIdx2 = new StringIndexer()
              .setInputCols(Array("genre", "rating"))
              .setOutputCols(Array("genreIdx", "ratingIdx"))

movieIdx2.fit(movie_data2)
        .transform(movie_data2)
        .orderBy('genreIdx)
        .show()
```

```
+---+------+------+--------+---------+
| id| genre|rating|genreIdx|ratingIdx|
+---+------+------+--------+---------+
|  3|Comedy| NC-17|     0.0|      1.0|
|  1|Comedy|     G|     0.0|      0.0|
|  2|Action|    PG|     1.0|      2.0|
|  4|Horror| PG-13|     2.0|      3.0|
+---+------+------+--------+---------+
```

In the scenario where a particular categorical value exists in the training dataset but doesn't in the test dataset. By default, the StringIndexer estimator throws an error to indicate such a scenario. It provides two additional ways to deal with this situation.

- **skip**: filter out rows with invalid data

- **keep**: put the invalid data in a special additional bucket

You can state how you want the StringIndexer estimator to handle this scenario by specifying the following parameters to the setHandleInvalid function: keep, skip, error.

Another useful estimator when working with categorical values is the OneHotEncoderEstimator, which encodes the index of a categorical value into a binary vector. The OneHotEncoder transformer has been deprecated starting with Spark version 2.3.0 due to its limitation in handling unknown categories. This estimator is often used in conjunction with the StringIndexer estimator, where the output of StringIndexer becomes the input of this estimator. Listing 8-13 demonstrates the usage of both estimators.

Listing 8-13. OneHotEncoderEstimator Consumes the Output of the StringIndexer Estimator

```
import org.apache.spark.ml.feature.OneHotEncoderEstimator

// the input column genreIdx is the output column of StringIndex in
   listing 8-9
val oneHotEncoderEst = new OneHotEncoderEstimator().setInputCols(
                            Array("genreIdx"))
                  .setOutputCols(Array("genreIdxVector"))
```

```
// fit the indexedMovie data produced in listing 8-10
val oneHotEncoderModel = oneHotEncoderEst.fit(indexedMovie)

val oneHotEncVect = oneHotEncoderModel.transform(indexedMovie)

oneHotEncVect.orderBy("genre").show()

+---+--------+---------+-------------------+
|id |  genre |  genreIdx|      genreIdxVector|
+---+--------+---------+-------------------+
| 5 | Action |   1.0   |     (2,[1],[1.0]) |
| 2 | Action |   1.0   |     (2,[1],[1.0]) |
| 3 | Comedy |   2.0   |     (2,[],[])     |
| 6 | Comedy |   2.0   |     (2,[],[])     |
| 1 | Comedy |   2.0   |     (2,[],[])     |
| 4 | Horror |   0.0   |     (2,[0],[1.0])|
+---+--------+---------+-------------------+
```

The Word2Vec estimator is useful when working with free text. It stands for *words to vector*. This estimator utilizes a well-known word embeddings technique that converts word tokens into numeric vector representations such that semantically similar words are mapped to nearby points. The intuition behind this technique is that similar words tend to occur together and have similar contexts. In other words, when two different words have very similar neighboring words, then they are probably quite similar in meaning or are related. This technique has proven effective in several natural language processing applications such as word analogies, word similarities, entity recognition, and machine translation.

The Word2Vec estimator has a few configurations, and the appropriate values need to be provided to control the output. Table 8-6 describes the configurations.

Table 8-6. *Word2Vec Configurations*

Name	Default Value	Description
vectorSize	100	The size of the output vector.
windowSize	5	The number of words to be used as the context.
minCount	5	The minimum number of times a token must appear to be included in the output.
maxSentenceLength	1000	Interaction between other terms to create new features. Multiplication is used for numeric values and binarized for categorical values.

Listing 8-14 demonstrates how to use the Word2Vec estimator and shows how to find similar words.

Listing 8-14. Use Word2Vec Estimator to Compute Word Embeddings and Find Similar Words

```
import org.apache.spark.ml.feature.Word2Vec

val documentDF = spark.createDataFrame(Seq(
              "Unified data analytics engine Spark".split(" "),
              "People use Hive for data analytics".split(" "),
              "MapReduce is not fading away".split(" "))
                    .map(Tuple1.apply)).toDF("word")

val word2Vec = new Word2Vec().setInputCol("word")
                      .setOutputCol("feature")
                      .setVectorSize(3)
                      .setMinCount(0)

val model = word2Vec.fit(documentDF)
val result = model.transform(documentDF)

result.show(false)
```

```
+-----------------------------------------+--------------------------------------------------------------------+
|word                                     |feature                                                             |
+-----------------------------------------+--------------------------------------------------------------------+
|[Unified, data, analytics, engine, Spark]|[-0.04857720620930195,-0.039790508151054386,-0.0047628857195377355] |
|[People, use, Hive, for, data, analytics]|[-0.019269779634972412,-0.0019863341003656387,0.04896292210711787]  |
|[MapReduce, is, not, fading, away]       |[0.09048619866371155,0.02390633299946785,0.004982998222112656]     |
+-----------------------------------------+--------------------------------------------------------------------+
```

```
// find similar words to Spark, the result shows both Hive and MapReduce
   are similar.
model.findSynonyms("Spark", 3).show
```

```
+---------------+----------------------------+
|           word|                  similarity|
+---------------+----------------------------+
|         engine|          0.9133241772651672|
|      MapReduce|          0.7623026967048645|
|           Hive|          0.7179173827171326|
+---------------+----------------------------+
```

```
// find similar words to Hive, the result shows Spark is similar
model.findSynonyms("Hive", 3).show
```

```
+---------+----------------------------+
|     word|                  similarity|
+---------+----------------------------+
|    Spark|          0.7179174423217773|
|   fading|          0.5859972238540649|
|   engine|          0.43200281262397766|
+---------+----------------------------+
```

The next estimators are about normalizing and standardizing numeric data. The reason for using these estimators is to ensure the learning algorithms that use distance as a measure don't place more weight on a feature with large values than another feature with smaller values.

Normalizing numeric data is the process of mapping its original range into a range from zero to one. This is especially helpful when observations have more than one attribute with different ranges. For example, say you have an employee's salary and height. The value for salary is in the thousands. The value for height is in a single digit. This is what the MinMaxScaler estimator is designed for. It linearly rescales each feature (column) individually to a common range of min and max values using the column summary statistics. For example, if the minimum value is 0.0 and the maximum value is 3.0, all the values fall within that range. Listing 8-15 provides an example of working with MinMaxScaler using the employee_data with salary and height information. The magnitude between the values of these two features is pretty big, but after running through the MinMaxScaler, that is not the case anymore.

Listing 8-15. Use MinMaxScaler to Rescale Features

```
import org.apache.spark.ml.feature.MinMaxScaler
import org.apache.spark.ml.linalg.Vectors

val employee_data = spark.createDataFrame(Seq(
                        (1, Vectors.dense(125400, 5.3)),
                        (2, Vectors.dense(179100, 6.9)),
                        (3, Vectors.dense(154770, 5.2)),
                        (4, Vectors.dense(199650, 4.11))))
                    .toDF("empId", "features")

val minMaxScaler = new MinMaxScaler().setMin(0.0)
                                .setMax(5.0)
                                .setInputCol("features")
                                .setOutputCol("sFeatures")

val scalerModel = minMaxScaler.fit(employee_data)

val scaledData = scalerModel.transform(employee_data)

println(s"Features scaled to range:
        [${minMaxScaler.getMin}, ${minMaxScaler.getMax}]")
Features scaled to range: [0.0, 5.0]

scaledData.select("features", "sFeatures").show(false)

+-------------------+----------------------------------------+
|      features     | scaledFeatures                         |
+-------------------+----------------------------------------+
|     [125400.0,5.3] | [0.0,2.1326164874551963]               |
|     [179100.0,6.9] | [3.616161616161616,5.0]                |
```

```
|      [154770.0,5.2]  | [1.9777777777777779,1.9534050179211468]  |
|      [199650.0,4.11] | [5.0,0.0]                                |
+-------------------+--------------------------------------------+
```

Besides the numeric data normalization, another operation that is often used for working with numeric data is *standardization*. This operation is especially applicable when the numeric data has a bell-shaped curve distribution. The standardization operation can help shift the data to a normalized form, where data is in a range of negative 1 and –1 with a mean of 0. The reason for doing this is to help certain ML algorithms learn better when the data has distribution around the mean of zero. The StandardScaler estimator is designed for the standardization operation. Listing 8-16 uses the same input dataset as in Listing 8-14. The output shows the values of the features are now centered around 0 and with one unit of standard deviation.

Listing 8-16. Use StandardScaler to Standard the Features Around the Mean of Zero

```scala
import org.apache.spark.ml.feature.StandardScaler
import org.apache.spark.ml.linalg.Vectors

val employee_data = spark.createDataFrame(Seq(
                            (1, Vectors.dense(125400, 5.3)),
                            (2, Vectors.dense(179100, 6.9)),
                            (3, Vectors.dense(154770, 5.2)),
                            (4, Vectors.dense(199650, 4.11))))
                    .toDF("empId", "features")

// set the unit standard deviation to true and center around the mean
val standardScaler = new StandardScaler().setWithStd(true)
                                    .setWithMean(true)
                                    .setInputCol("features")
                                    .setOutputCol("sFeatures")

val standardMode = standardScaler.fit(employee_data)

val standardData = standardMode.transform(employee_data)

standardData.show(false)
```

```
+-----+--------------+-------------------------------------------+
|empId| feature      |                sFeatures                  |
+-----+--------------+-------------------------------------------+
|   1 |[125400.0,5.3]|[-1.2290717420781212,-0.06743742573177587] |
|   2 |[179100.0,6.9]|  [0.4490658767775897,1.3248191055048935]  |
|   3 |[154770.0,5.2]|[-0.3112523404805006,-0.15445345893406737] |
|   4 |[199650.0,4.1]|  [1.091258205781032,-1.102928220839048]   |
+-----+--------------+-------------------------------------------+
```

There are many more estimators available in MLlib to perform numerous data transformations and mappings. They all follow a standard abstraction that fits the input data and produces an instance of Model. These examples are meant to illustrate how to work with these estimators. Examples of the second kind of estimators, ML algorithms, are covered in the following sections.

Pipeline

In machine learning, it is common to run a sequence of steps to clean and transform data, then train one or more ML algorithms to learn from the data, and finally tune the model to achieve the optimal model performance. The pipeline abstraction in MLlib is designed to make this workflow easier to develop and maintain. From the technical perspective, MLlib has a Pipeline class that is designed to manage a series of stages. Each one is represented by the PipelineStage class, either a transformer or an estimator. The Pipeline abstraction is a type of estimator.

The first step in setting up a pipeline is to create a collection of stages, create an instance of the Pipeline class, and configure it with an array of stages. The Pipeline class runs those stages in the specified order. If a stage is a transformer, then the transform() function is called. If a stage is an estimator, the fit() function is called to produce a transformer.

Let's walk through a small workflow example of processing text using transformers and estimators. The small pipeline depicted in Figure 8-10 consists of two transformers and one estimator. When the Pipeline.fit() function is called, the raw text of the input DataFrame passes to the Tokenizer transformer, and its output is passed into the HashingTF transformer, which converts the words into features. The Pipeline class recognizes LogisticRegression as an estimator, invoking the fit function with the computed features to produce a logistic regression model.

The code for the `Pipeline` is depicted in Figure 8-10 is in Listing 8-17. A `Pipeline` abstraction is an estimator. So once an instance of `Pipeline` is created and configured, the `fit()` function must be called with the training data as the input to trigger the execution of the stages. The output is an instance of `PipelineModel`, which is a type of transformer. At this point, you can pass the test data into the `transform()` function to perform predictions.

MLlib provides an ML persistence feature to make it easy to save a pipeline or a model to disk and load it later to perform predictions. The nice thing about the persistence feature is that it is designed to save the information in a language-neutral format. So, when a pipeline or model is persisted in Scala, it can be read back in a different language, such as Java or Python.

Many real-life production pipelines consist of many stages. When the number of stages is large, it is difficult to understand the flow and maintain. MLlib `Pipeline` abstraction can help with these challenges. Another key point to note is that both `Pipelines` and `PipelineModels` are designed to ensure the training and test data flow through the identical feature processing steps. One common mistake in machine learning is not processing the training and test data consistently, which creates a discrepancy in the model evaluation results.

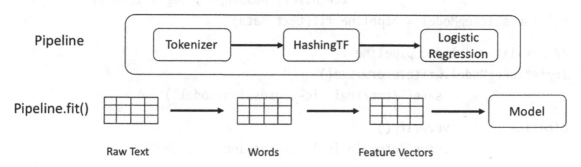

Figure 8-10. *Example of a small pipeline*

Listing 8-17. Using Pipepline to Small a Small Workflow

```
import org.apache.spark.ml.{Pipeline, PipelineModel}
import org.apache.spark.ml.classification.LogisticRegression
import org.apache.spark.ml.feature.{HashingTF, Tokenizer}
```

```scala
val text_data = spark.createDataFrame(Seq(
        (1, "Spark is a unified data analytics engine", 0.0),
        (2, "Spark is cool and it is fun to work with Spark", 0.0),
        (3, "There is a lot of exciting sessions at upcoming Spark
            summit", 0.0),
        (4, "signup to win a million dollars", 0.0)  ))
                    .toDF("id", "line", "label")

val tokenizer = new Tokenizer().setInputCol("line")
                                .setOutputCol("words")

val hashingTF = new HashingTF()
                    .setInputCol(tokenizer.getOutputCol)
                    .setOutputCol("features")
                    .setNumFeatures(4096)

val logisticReg = new LogisticRegression().setMaxIter(5)
                                            .setRegParam(0.01)

val pipeline = new Pipeline().setStages(Array(
                        tokenizer, hashingTF, logisticReg))
val logisticRegModel = pipeline.fit(text_data)

// persist model and pipeline
logisticRegModel.write.overwrite()
            .save("/tmp/logistic-regression-model")

pipeline.write.overwrite()
            .save("/tmp/logistic-regression-pipeline")

// load model and pipeline
val prevModel = PipelineModel.load("/tmp/spark-logistic-regression-model")

val prevPipeline = Pipeline.load("/tmp/logistic-regression-pipeline")
```

Pipeline Persistence: Saving and Loading

Once a model is trained and evaluated, you can save that model or the pipeline that trained that model to further evaluate your model with additional datasets on future days or after the Spark cluster was restarted. The later approach is preferred because it remembers the model type; otherwise, you must specify it during the loading step.

The main benefit of persisting your model is to save time and skip the training step, which might take many hours to complete.

Model Tuning

The model tuning step aims to train a model using a set of parameters to achieve the optimal model performance to meet the objectives defined in the first step of the ML development process. This step is usually tedious, repetitive, and time-consuming because it requires experimenting with different ML algorithms or a few sets of parameters.

This section aims to describe a few tools MLlib provides to help with the laborious part of the model tuning step. It is not the intention of this section to show how to perform model tuning.

Before going into the details of the tools MLlib provides, it is important to understand the following terminologies.

- Model hyperparameters are

  - Configurations that govern the ML algorithm training process

  - Configurations that are external to the model and can't be learned from the training data

  - Configurations that the machine learning practitioners provide before the training process starts

  - Configurations that are tuned for a given machine learning task through an iterative manner

- Model parameters are

  - Properties that are not provided by the machine learn practitioners

  - Properties of the training data that are learned during the training process

- Properties that are optimized during the training process

- Properties of the model that perform predictions

Examples of model hyperparameters include the number of clusters in the k-means clustering algorithm, the amount of regularization applied in the logistic regression algorithm, and the learning rate.

Examples of the model parameters include the coefficients in a linear regression model or the branch locations in the decision tree model.

The two commonly used classes in MLlib to help with model tuning are `CrossValidator` and `TrainValidationSplit,` and both are of type `Estimator`. These classes are also known as *validators*, and they require the following input.

- The first input is what needs to be tuned—an ML algorithm or an instance of `Pipeline`. It must be a type of estimator.

- The second input is a set of parameters to use to tune the provided estimator. These parameters are also known as a *parameter grid* to search over to find the best model. A convenient utility called `ParagramGridBuilder` is available to use to build the parameter grid.

- The last input is an evaluator to evaluate the performance of a model based on the held-out test data. MLlib provides a specific evaluator for each machine learning task, which can produce one or more evaluation metrics for you to understand the model performance. The commonly used machine learning metrics are supported, such as root mean square error, precision, recall, and accuracy

At the high level, validators perform the following steps with the given inputs.

1. The input feature data is split into training and test dataset based on the specified ratio.

2. For each combination in the parameter grid, the given estimator is fitted with the training data and the parameter combination.

3. The specified evaluator evaluates the output model against the test data. The performance metric is recorded and compared.

4. The model that produces the best performance is returned along with the set of parameters that were used.

These steps are illustrated in Figure 8-11, making it easier to visualize what's going on in the validator.

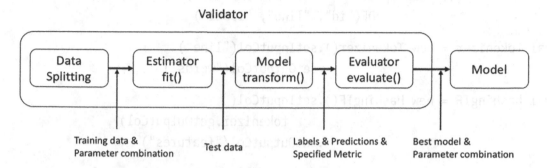

Validator

| Data Splitting | → | Estimator fit() | → | Model transform() | → | Evaluator evaluate() | → | Model |

Training data & Parameter combination   Test data   Labels & Predictions & Specified Metric   Best model & Parameter combination

Figure 8-11. *Inside a validator*

The TrainValidationSplit validator splits the given input data into the training and validation dataset based on the specified ratio and then trains and evaluates the dataset pair against each parameter combination. For example, if the given parameter set has six combinations, the given estimator is trained and evaluated size times, each time with a different parameter combination.

Listing 8-18 provides an example of using TrainValidationSplit to tune a linear regression estimator with a parameter grid of six parameter combinations. The focus of this example is TrainValidationSplit. There is an assumption that the feature engineering has already been done, and there is a column called features in the DataFrame.

Listing 8-18. Example of TrainValidationSplit

```
import org.apache.spark.ml.{Pipeline, PipelineModel}
import org.apache.spark.ml.classification.LogisticRegression
import org.apache.spark.ml.feature.{HashingTF, Tokenizer}
import org.apache.spark.ml.tuning.{ParamGridBuilder, TrainValidationSplit}
import org.apache.spark.ml.evaluation.BinaryClassificationEvaluator

val text_data = spark.createDataFrame(Seq(
        (1, "Spark is a unified data analytics engine", 0.0),
        (2, "Spark is cool and it is fun to work with Spark",
            0.0),
```

```
              (3, "There is a lot of exciting sessions at upcoming
                    Spark summit", 0.0),
              (4, "signup to win a million dollars", 0.0)  ))
                    .toDF("id", "line", "label")

val tokenizer = new Tokenizer().setInputCol("line")
                              .setOutputCol("words")

val hashingTF = new HashingTF().setInputCol(
                                  tokenizer.getOutputCol)
                    .setOutputCol("features")

val logisticReg = new LogisticRegression().setMaxIter(5)

val pipeline = new Pipeline().setStages(
                    Array(tokenizer, hashingTF, logisticReg))

// the first parameter has 3 values and second parameter has 2 values,
// therefore the total parameter combinations is 6
val paramGrid = new ParamGridBuilder().addGrid(
                    hashingTF.numFeatures, Array(10, 100, 250))
            .addGrid(logisticReg.regParam, Array(0.1, 0.05))
            .build()

// setting up the validator with required inputs - estimator, evaluator,
parameter grid and train ratio
val trainValSplit = new TrainValidationSplit()
                    .setEstimator(pipeline)
                  .setEvaluator(
                      new BinaryClassificationEvaluator)
                  .setEstimatorParamMaps(paramGrid)
                  .setTrainRatio(0.8)

// train the linear regression estimator
val model = trainValidationSplit.fit(training)
```

The CrossValidator validator implements a widely known technique in the machine learning community to help with the model tuning step. This technique maximizes the amount of data for training and test by randomly dividing the observations into non-overlapping _k_ groups, or folds, of approximately the same size.

Each one is used only once. One fold is used for testing, and the remaining ones are used for training. This process is repeated k times, and each time the estimator is trained and evaluated against randomly divided training and test folds.

Figure 8-12 illustrates this process with k as the four folds. CrossValidator generates four training and test dataset pairs, and one-fourth of the data is for testing, and three-fourths of the data is for testing. It is important to select a reasonable k value so that each training and testing group is statistically representative of the available observation. Each fold has roughly the same amount of sample data.

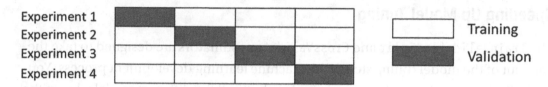

Figure 8-12. *K-fold example with k=4*

It is important to be aware of the long completion time when using this validator with a large number of parameter combinations. This is because each experiment described in Figure 8-12 is performed against each parameter combination. For example, if k is 4 and the number of parameter combinations is 6, then the total number of times the estimator is trained and evaluated is 24. Listing 8-17 replaces the TrainValidationSplit in Listing 8-15 with an instance of CrossValidator, and it is configured with 4 as the k value. In practice, the value for k is usually 10 or higher. Listing 8-17 ends up training and evaluating the estimator 25 times.

After the model with the best performance is identified, CrossValidator retrains or refits your model using the same set of parameters on the entire dataset. That's the reason the model in Listing 8-19 is trained a total of 25 times.

Listing 8-19. Example of CrossValidator

```
import org.apache.spark.ml.tuning.CrossValidator

val crossValidator = new CrossValidator()
                    .setEstimator(pipeline)
                    .setEvaluator(
                        new BinaryClassificationEvaluator)
```

```
                    .setEstimatorParamMaps(paramGrid)
                    .setNumFolds(4)

val model = crossValidator.fit(text_data)
```

If there is a need to study or analyze the intermediate models, the `CrossValidator` can retain them during the tuning process. All you need to do is specify a true value when calling the `setCollectSubmModels` function and then access the intermediate models by calling the `getCollectSubmModels()` function.

Speeding Up Model Tuning

The `TrainValidationSplit` and `CrossValidator` estimators are designed to take the pain out of the model tuning step in the machine learning development process. You might find that it takes a while to train and evaluate all the different models due to the different parameter combinations. The larger the number of parameter combinations, the more time it takes.

By default, the estimators are training and evaluating one model at a time in a sequential manner. To speed up this process, you might want to increase the parallelism to take advantage of your Spark cluster's compute and memory resources. This is done by setting the parallelism with a value of 2 or greater before initiating the model tuning process. As a general guideline from the Spark tuning guide, a value up to 10 is usually sufficient. Listing 8-20 sets the parallelism of crossValidator to 6.

Listing 8-20. Setting CrossValidator Parallelism to 6

```
crossValidator.setParallelism(6).fit(text_data)
```

Model Evaluators

To understand how well a model performs, you first need to know how to calculate and evaluate the model evaluation metrics. Each machine learning task uses a different set of metrics, and calculating them is tedious and using math. Luckily, MLlib provides a set of tools called an *evaluator* to calculate the metric so the validator can measure how well a fitted model does on test data. Table 8-7 lists out the different supported evaluators in MLlib, a subset of the supported metrics, and a short description.

Table 8-7. Supported Evaluators

Name	Supported Metrics	Description
RegressionEvaluator	rmse, mse, r2, mae, var	For regression task
BinaryClassificationEvaluator	areaUnderROC, areaUnderPR	For classification task with only two classes
MulticlassClassificationEvaluator	weightedPrecision, weightedRecall, etc	For classification task with only more than two classes
MultilabelClassificationEvaluator	subsetAccuracy, accuracy, hammingLos, recall, precisionByLabel, recallByLabel, f1MeasureByLabel	For multi-label classification task
RankingEvaluator	meanAveragePrecisionAtK, precisionAtK, ndcgAtK, recallAtK	For ranking tasks

Machine Learning Tasks in Action

This section brings together the concepts and tools described in this chapter and applies them to the following machine learning tasks: classification, regression, and recommendation. Working through the machine learning development process with real datasets makes it clearer how all the pieces fit together.

This section is not meant to be comprehensive in covering the hyperparameters of each machine learning algorithm, and the model tuning step is left as an exercise for the readers.

Classification

Classification is one of the most widely studied and used machine learning tasks due to its ability to help with solving many real-life classification-related problems. For example, is this a fraudulent credit card transaction? Is this email spam? Is this an image of a cat, a dog, or a bird?

There are three types of classification.

- **Binary classification**: This is where the label to predict has only two possible classes (for example, fraud or not fraud, conference paper is accepted or not, the tumor is benign or malignant).

- **Multiclass classification**: This is where the label to predict has more than two possible classes (for example, whether an image is a dog, cat, or bird).

- **Multilabel classification**: This is where each observation can belong to more than one class. Movie genres are a good example of this. A movie can be classified as both action and comedy. MLlib doesn't natively support this type of classification.

MLlib provides a few machine learning algorithms for the classification tasks.

- Logistic regression

- Decision tree

- Random forest

- Gradient-boosted tree

- Linear support-vector machine

- One-vs-Rest

- Naïve Bayes

Model Hyperparameters

The logistic regression algorithm is used in this example. The following is a subset of its model hyperparameters. Each model hyperparameter has a default value.

- `family`: The possible values are `auto`, `binomial`, and `multinomial`. The default value is `auto`, which means the algorithm automatically selects the family to be either `binomial` or `multinomial` based on the values in the label column. `binomial` is for binary classification. `multinomial` is for the multiclass classification.

- `regParam`: This is the regularization parameter to control the overfitting. The default value is 0.0.

Example

Listing 8-21 tries to predict which *Titanic* passengers survived the tragedy. This is a binary classification machine learning problem. The example uses the logistic regression algorithm. The information and the data are available at www.kaggle.com/c/titanic. The data is in CSV format, and there are two files: train.csv and test.csv. The train.csv file contains the label column.

The provided data contains many interesting features; however, Listing 8-21 uses only age, gender, and ticket class as features.

Listing 8-21. Use Logistic Regression Algorithm to Predict the Survival of Titanic Passengers

```
import org.apache.spark.ml.Pipeline
import org.apache.spark.ml.feature.StringIndexer
import org.apache.spark.ml.feature.VectorAssembler
import org.apache.spark.ml.classification.LogisticRegression
import org.apache.spark.ml.evaluation.BinaryClassificationEvaluator

val titanic_data = spark.read.format("csv")
                        .option("header", "true")
                        .option("inferSchema","true")
                        .load("/<folder>/train.csv")

// explore the schema
titanic_data.printSchema
  |-- PassengerId: integer (nullable = true)
  |-- Survived: integer (nullable = true)
  |-- Pclass: integer (nullable = true)
  |-- Name: string (nullable = true)
  |-- Sex: string (nullable = true)
  |-- Age: double (nullable = true)
  |-- SibSp: integer (nullable = true)
  |-- Parch: integer (nullable = true)
  |-- Ticket: string (nullable = true)
  |-- Fare: double (nullable = true)
  |-- Cabin: string (nullable = true)
  |-- Embarked: string (nullable = true)
```

```
// to start out with, we will use only three features
// filter out rows where age is null
val titanic_data1 = titanic_data.select('Survived.as("label"),
                      'Pclass.as("ticket_class"),
                      'Sex.as("gender"), 'Age.as("age"))
                            .filter('age.isNotNull)

// split the data into training and test with 80% and 20% split
val Array(training, test) = titanic_data1.randomSplit(
                                    Array(0.8, 0.2))

println(s"training count: ${training.count}, test count:
                        ${test.count}")

// estimator:  to convert gender string to numbers
val genderIndxr = new StringIndexer().setInputCol("gender")
                                    .setOutputCol("genderIdx")

// transformer: assemble the features into a vector
val assembler = new VectorAssembler().setInputCols(
                    Array("ticket_class", "genderIdx", "age"))
                                .setOutputCol("features")

// estimator: the algorithm
val logisticRegression = new LogisticRegression()
                                    .setFamily("binomial")

// set up the pipeline with three stages
val pipeline = new Pipeline().setStages(Array(genderIndxr,
                                assembler, logisticRegression))

// train the algorithm with the training data
val model = pipeline.fit(training)

// perform the predictions
val predictions = model.transform(test)

// perform the evaluation of the model performance, the default metric is
the area under the ROC
val evaluator = new BinaryClassificationEvaluator()
```

```
evaluator.evaluate(predictions)
res10: Double = 0.8746657754010692

evaluator.getMetricName
res11: String = areaUnderROC
```

The metric produced by BinaryClassificationEvaluator has a value of 0.87, which is a decent performance for using three features. However, this example doesn't explore the various hyperparameters and training parameters. I highly recommend that you experiment with the various hyperparameters to see if your model can perform better than 0.87.

Regression

Another popular machine learning task is called regression, which is designed to predict a real number or continuous value. For example, you want to predict the sales revenue for the next quarter, or the income of a population, or the amount of rain in a certain region of the world.

MLlib provides the following machine learning algorithms for regression tasks.

- Linear regression

- Generalized linear regression

- Decision trees

- Random forest

- Gradient-boosted trees

- Isotonic regression

Model Hyperparameters

The following example uses linear regression with the following hyperparameters.

- regParam: This regularization parameter controls the overfitting. The default value is 0.0.

- fitIntercept: This parameter determines whether to fit the intercept. The default value is true.

Example

Listing 8-22 tries to predict a house price based on a set of features about the houses. The dataset is available at www.kaggle.com/c/house-prices-advanced-regression-techniques/data. The data is provided in CSV format, and there are two files, train.csv, and test.csv. The label column in the train.csv file is called SalePrice.

The provided data contains many interesting features; however, Listing 8-22 uses only a subset of them.

Listing 8-22. Use Linear Regression Algorithm to Predict House Price

```
import org.apache.spark.sql.functions._
import org.apache.spark.ml.Pipeline
import org.apache.spark.ml.feature.StringIndexer
import org.apache.spark.ml.feature.VectorAssembler
import org.apache.spark.ml.regression.LinearRegression
import org.apache.spark.ml.feature.RFormula
import org.apache.spark.ml.evaluation.RegressionEvaluator
import org.apache.spark.mllib.evaluation.RegressionMetrics

val house_data = spark.read.format("csv")
                    .option("header", "true")
                    .option("inferSchema","true")
                    .load("<path>/train.csv")

// select columns to use as features
val cols = Seq[String]("SalePrice", "LotArea",  "RoofStyle",
                    "Heating", "1stFlrSF", "2ndFlrSF",
                    "BedroomAbvGr", "KitchenAbvGr",
                    "GarageCars", "TotRmsAbvGrd",
                    "YearBuilt")

val colNames = cols.map(n => col(n))

// select only needed columns
val skinny_house_data = house_data.select(colNames:_*)

// create a new column called "TotalSF" by adding the value of "1stFlrSF"
   and "2ndFlrSF" columns
```

```
// cast the "SalePrice" column to double
val skinny_house_data1 = skinny_house_data.withColumn("TotalSF",
                         col("1stFlrSF") + col("2ndFlrSF"))
                    .drop("1stFlrSF", "2ndFlrSF")
                    .withColumn("SalePrice",
                         $"SalePrice".cast("double"))

// examine the statistics of the label column called "SalePrice"
skinny_house_data1.describe("SalePrice").show
```

```
+-----------+-------------------+
|    summary|          SalePrice|
+-----------+-------------------+
|      count|               1460|
|       mean| 180921.19589041095|
|     stddev|  79442.50288288663|
|        min|            34900.0|
|        max|           755000.0|
+-----------+-------------------+
```

```
// create estimators and transformers to setup a pipeline

// set the invalid categorical value handling policy to skip to avoid error
// at evaluation time
val roofStyleIndxr = new StringIndexer()
                         .setInputCol("RoofStyle")
                         .setOutputCol("RoofStyleIdx")
                         .setHandleInvalid("skip")

val heatingIndxr = new StringIndexer()
                         .setInputCol("Heating")
                         .setOutputCol("HeatingIdx")
                         .setHandleInvalid("skip")

val linearReg = new LinearRegression().setLabelCol("SalePrice")

// assembler to assemble the features into a feature vector
val assembler = new VectorAssembler().setInputCols(
                    Array("LotArea", "RoofStyleIdx",
```

```
                              "HeatingIdx", "LotArea",
                              "BedroomAbvGr", "KitchenAbvGr",
                              "GarageCars", "TotRmsAbvGrd",
                              "YearBuilt", "TotalSF"))
                                    .setOutputCol("features")

// setup the pipeline
val pipeline = new Pipeline().setStages(Array(roofStyleIndxr,
                    heatingIndxr, assembler, linearReg))

// split the data into training and test pair
val Array(training, test) = skinny_house_data1.randomSplit(
                                    Array(0.8, 0.2))

// train the pipeline
val model = pipeline.fit(training)

// perform prediction
val predictions = model.transform(test)

val evaluator = new RegressionEvaluator().setLabelCol("SalePrice")

setPredictionCol("prediction")

setMetricName("rmse")

val rmse = evaluator.evaluate(predictions)
rmse: Double = 37579.253919082395
```

RMSE stands for the root-mean-square error. In this case, the RMSE value is around $37,000, which indicates there is a lot of room for improvement.

Recommendation

Recommender system is one of the most intuitive and well-known machine learning applications. Maybe that is the case because almost everyone has seen examples of recommender systems in action on popular websites such as Amazon and Netflix. Almost every single popular website or Internet e-commerce company has one or more recommender systems on their site. Popular examples of recommender systems are songs you may like on Spotify, people you want to follow on Twitter, courses you may like on Coursera or Udacity. The benefits recommender systems bring mutually beneficial to the company's users and itself. Users are delighted to find or discover items that they like without spending too much effort. Companies are happy due to the increased user engagement, loyalty, and their bottom line. If a recommender system performs well, it is a win-win situation.

The common approaches to building recommender systems include content-based filtering, collaborative filtering, and a hybrid of the two. The first approach requires collecting information about the items being recommended and the profile of each user. The second approach requires collecting only user activities or behavior via explicit or implicit means. Examples of explicit behavior include rating a movie or an item on Amazon. Examples of implicit behavior including viewing the movie trailer or description. The intuition behind the second approach is the "wisdom of the crowd" concept, where people who agreed in the past tend to agree in the future.

This section focuses on the collaborative filter approach and one of the popular algorithms for this approach is called ALS, which stands for *alternate-least-square*. The only input this algorithm needs are the user-item rating matrix, which discovers user preferences and item properties through *matrix factorization*. Once these two pieces of information are found, they predict the user's preference on items not seen before. MLlib has an implementation of the ALS algorithm.

Model Hyperparameters

The ALS algorithm implementation in MLlib has a few important hyperparameters that you need to be aware of. The following section contains just a subset. Please consult the documentation at `https://spark.apache.org/docs/latest/ml-collaborative-filtering.html`.

- `rank`: This parameter specifies the number of latent factors or properties about users and items learned during the training process. An optimal value for rank is usually determined by experimentation and an intuition about the number of properties needed to accurately describe an item. The default value is 10.

- `regParam`: The amount of regularization to deal with overfitting. An optimal value for this parameter is usually determined by experimentation. The default is 0.1

- implicitPrefs: ALS algorithm supports both explicit and implicit user activities or behavior. This parameter tells which one the input data represents. The default is false, meaning the activities or behavior are explicit.

Example

This example builds a movie recommender system using the movie ratings dataset at https://grouplens.org/datasets/movielens/. The specific dataset is the latest MovieLens 100K dataset at http://files.grouplens.org/datasets/movielens/ml-latest-small.zip. This dataset contains roughly about 100,000 ratings by 700 users across 9000 movies. There are four files included in the zip file: links.csv, movies.csv, ratings.csv, and tags.csv. Each row in the ratings.csv file represents one rating of one movie by one user. It is in this format: userId, movieId, rating, timestamp. The rating is on a scale from 0 to 5 with half-star increments.

Listing 8-23 trains the ALS algorithm with one set of parameters and then evaluates the model performance based on the RMSE metric. In addition, it calls a few interesting provided APIs in the ALSModel class to get recommendations for movies and users.

Listing 8-23. Building a Recommender System Using ALS Algorithm Implementation in MLlib

```
import org.apache.spark.mllib.evaluation.RankingMetrics
import org.apache.spark.ml.evaluation.RegressionEvaluator
import org.apache.spark.ml.recommendation.ALS
import org.apache.spark.ml.tuning.{ParamGridBuilder, CrossValidator}
import org.apache.spark.sql.functions._

// we don't need the timestamp column, so drop it immediately
val ratingsDF = spark.read.option("header", "true")
                          .option("inferSchema", "true")
                          .csv("<path>/ratings.csv")
                          .drop("timestamp")

// quick check on the number of ratings
ratingsDF.count

res14: Long = 100004
```

```
// quick check who are the active movie raters
val ratingsByUserDF = ratingsDF.groupBy("userId").count()

ratingsByUserDF.orderBy($"count".desc).show(10)
```

```
+--------+-------+
| userId|  count|
+--------+-------+
|     547|   2391|
|     564|   1868|
|     624|   1735|
|      15|   1700|
|      73|   1610|
|     452|   1340|
|     468|   1291|
|     380|   1063|
|     311|   1019|
|      30|   1011|
+--------+-------+
```

```
println("# of rated movies: " +ratingsDF.select("movieId").distinct().
count)
# of rated movies: 9066
```

```
println("# of users: " + ratingsByUserDF.count)
# of users: 671
```

```
// analyze the movies largest number of ratings
val ratingsByMovieDF = ratingsDF.groupBy("movieId").count()
ratingsByMovieDF.orderBy($"count".desc).show(10)
```

```
+----------+-------+
| movieId|  count|
+----------+-------+
|     356|    341|
|     296|    324|
|     318|    311|
|     593|    304|
```

```
|       260|    291|
|       480|    274|
|      2571|    259|
|         1|    247|
|       527|    244|
|       589|    237|
+----------+-------+
```

```scala
// prepare data for training and testing
val Array(trainingData, testData) = ratingsByUserDF.randomSplit(
Array(0.8, 0.2))

// setting up an instance of ALS
val als = new ALS().setRank(12)
                   .setMaxIter(10)
                   .setRegParam(0.03)
                   .setUserCol("userId")
                   .setItemCol("movieId")
                   .setRatingCol("rating")

// train the model
val model = als.fit(trainingData)

// perform predictions
val predictions = model.transform(testData).na.drop

// setup an evaluator to calculate the RMSE metric
val evaluator = new RegressionEvaluator().setMetricName("rmse")
                                         .setLabelCol("rating")
                                         .setPredictionCol
                                         ("prediction")

val rmse = evaluator.evaluate(predictions)
println(s"Root-mean-square error = $rmse")
Root-mean-square error = 1.06027809686058
```

The ALSModel class provides two sets of useful functions to perform recommendations. The first set recommends the top n items to all users or a specific set of users. The second set is for recommending top *n* users to all items or a specific set of items. Listing 8-24 provides an example of calling these functions.

Listing 8-24. Using ALSModel to Perform Recommendations

```
// recommend the top 5 movies for all users
model.recommendForAllUsers(5).show(false)

// active raters
val activeMovieRaters = Seq((547), (564), (624), (15),
                                    (73)).toDF("userId")

model.recommendForUserSubset(activeMovieRaters, 5).show(false)
```

```
+------+--------------------------------------------------------------------------------------------------+
|userId|                recommendations                                                                   |
+------+--------------------------------------------------------------------------------------------------+
|  15  | [[363, 5.4706035],   [422, 5.4109325],  [1192, 5.3407555], [1030, 5.329553],  [2467, 5.214414]]  |
|  547 | [[1298, 5.752393],   [1235, 5.4936843], [994, 5.426885],   [926, 5.28749],    [3910, 5.2009006]] |
|  564 | [[121231, 6.199452], [2454, 5.4714866], [3569, 5.4276495], [1096, 5.4212027], [1292, 5.4203687]] |
|  624 | [[1960, 5.4001703],  [1411, 5.2505665], [3083, 5.1079946], [3030, 5.0170803], [132333, 5.0165534]]|
|  73  | [[2068, 5.0426316],  [5244, 5.004793],  [923, 4.992707],   [85342, 4.979018], [1411, 4.9703207]]  |
+------+--------------------------------------------------------------------------------------------------+
```

```
// recommend top 3 users for each movie
val recMovies = model.recommendForAllItems(3)
```

```scala
// read in movies dataset so we can see the movie title
val moviesDF = spark.read.option("header", "true")
                        .option("inferSchema", "true")
                        .csv("<path>/movies.csv")

val recMoviesWithInfoDF = recMovies.join(moviesDF, "movieId")


recMoviesWithInfoDF.select("movieId", "title", "recommendations")
                .show(5, false)
```

```
+--------+----------------------------+------------------------------------------------------------+
| movieId| title                      | recommendations                                            |
+--------+----------------------------+------------------------------------------------------------+
|   1580 | Men in Black (a.k.a. MIB) (1997) | [[46, 5.6861496],  [113, 5.6780157], [145, 5.3410296]] |
|   5300 | 3:10 to Yuma (1957)        | [[545, 5.475599], [354, 5.2230153], [257, 5.0623646]] |
|   6620 | American Splendor (2003)   | [[156, 5.9004226], [83, 5.699677],  [112, 5.6194253]] |
|   7340 | Just One of the Guys (1985) | [[621, 4.5778027], [451, 3.9995837], [565, 3.6733315]] |
|  32460 | Knockin' on Heaven's Door (1997) | [[565, 5.5728054], [298, 5.00507],  [476, 4.805148]] |
+--------+----------------------------+------------------------------------------------------------+
```

```scala
// top rated movies
val topRatedMovies = Seq((356), (296), (318),
                        (593)).toDF("movieId")

// recommend top 3 users per movie in topRatedMovies
val recUsers =  model.recommendForItemSubset(topRatedMovies, 3)

recUsers.join(moviesDF, "movieId")
        .select("movieId", "title", "recommendations")
        .show(false)
```

```
+----------+-----------------------------------+-------------------------------------------------------+
| movieId| title                               | recommendations                                       |
+----------+-----------------------------------+-------------------------------------------------------+
| 296    | Pulp Fiction (1994)                 | [[4, 5.8505774],   [473, 5.81865],   [631, 5.588397]] |
| 593    | Silence of the Lambs, The (1991)    | [[153, 5.839533],  [586, 5.8279104], [473, 5.5933723]]|
| 318    | Shawshank Redemption, The (1994)    | [[112, 5.8578305], [656, 5.8488774], [473, 5.795221]] |
| 356    | Forrest Gump (1994)                 | [[464, 5.6555476], [58, 5.6497917],  [656, 5.625555]] |
+----------+-----------------------------------+-------------------------------------------------------+
```

In Listing 8-24, an instance of the ALS algorithm was trained with one set of parameters, and the RSME is about 1.06. Let's try retraining that instance of the ALS algorithm with a set of parameter combinations using the CrossValidator to see if you can lower the RSME value.

Listing 8-25 sets up a search grid for two hyperparameters with the total of four parameter combinations, and a CrossValidator with three folds. This means the ALS algorithm is trained and evaluated 12 times, and therefore it takes a minute or two to complete.

Listing 8-25. Use CrossValidator to Tune the ALS Model

```
val paramGrid = new ParamGridBuilder()
                    .addGrid(als.regParam,Array(0.05, 0.15))
                    .addGrid(als.rank, Array(12,20))
                    .build

val crossValidator = new CrossValidator().setEstimator(als)
                    .setEvaluator(evaluator)
                    .setEstimatorParamMaps(paramGrid)
                    .setNumFolds(3)
```

```
// print out the 4 hyperparameter combinations
crossValidator.getEstimatorParamMaps.foreach(println)
{
        als_d2ec698bdd1a-rank: 12,
        als_d2ec698bdd1a-regParam: 0.05
}
{
        als_d2ec698bdd1a-rank: 20,
        als_d2ec698bdd1a-regParam: 0.05
}
{
        als_d2ec698bdd1a-rank: 12,
        als_d2ec698bdd1a-regParam: 0.15
}
{
        als_d2ec698bdd1a-rank: 20,
        als_d2ec698bdd1a-regParam: 0.15
}

// this will take a while to run through more than 10 experiments
val cvModel = crossValidator.fit(trainingData)

// perform the predictions and drop the
val predictions2 = cvModel.transform(testData).na.drop

val evaluator2 = new RegressionEvaluator()
                                .setMetricName("rmse")
                                .setLabelCol("rating")
                                .setPredictionCol("prediction")

val rmse2 = evaluator2.evaluate(predictions2)
rmse2: Double = 0.9881840432547675
```

You have successfully lowered the RMSE by leveraging the CrossValidator to help with tuning the model. It may take a while to train the best model, but MLlib makes it easy to experiment with a set of parameter combinations.

Deep Learning Pipeline

This chapter would be incomplete if there is no reference to the deep learning topic, one of the hottest topics in the artificial intelligence and machine learning landscapes. There are already many resources available in the form of books, blogs, courses, and research papers to explain every aspect of deep learning. In terms of technology, there are many innovations from the open source community, universities and large companies like Google, Facebook, Microsoft, and others to come up with deep learning frameworks and best practices. Here is the current list of Deep Learning frameworks.

- TensorFlow is open source framework created by Google.

- PyTorch is an open source deep learning framework developed by Facebook.

- MXNet is a deep learning framework developed by a group of universities and companies.

- Caffe is a deep learning framework developed by UC Berkeley.

- CNTK is an open source deep learning framework developed by Microsoft.

- Theano is another open deep learning framework developed by the University of Montreal.

- BigDL is an open source deep learning framework developed by Intel.

On Apache Spark's side, Databricks is driving the effort of developing a project called Deep Learning Pipelines. It is not another deep learning framework, but rather it is designed to work on top of the existing popular deep learning frameworks. In the spirit of Spark and MLlib, the Deep Learning Pipelines project provides high-level and easy-to-use APIs for building scalable Deep Learning applications in Python using Apache Spark. This project is currently being developed outside of the Apache Spark open source project, and eventually, it will be incorporated into the main trunk. At the time of this writing, the Deep Learning Pipelines project provides the following features.

- Common deep learning use cases can be implemented in just a few lines of code.

- Working with images in Spark

- Apply pre-trained deep learning models for scalable predictions

- The ability to do transfer learning, which adapts a model trained for a similar task to the current ask

- Distributed hyperparameter tuning

- Make it easy to expose deep learning models so others can use them as a function in SQL to make predictions

More information about the exciting Deep Learning Pipelines project is at `https://github.com/databricks/spark-deep-learning`.

Summary

The adoption of artificial intelligence and machine learning is steadily increasing, and there are many exciting breakthroughs in the coming years. Building on top of the strong foundation of Spark, the MLlib component is designed to help build intelligent applications in an easy and scalable manner.

- Artificial intelligence is a broad field whose goal is to make machines seem like they have intelligence. Machine learning is one of the subfields; it focuses on teaching machines to learn by training them with data.

- Building machine learning applications consists of a sequence of steps, and it is highly iterative.

- The Spark MLlib component consists of tools and abstractions for feature engineering, constructing, evaluating, and tuning machine learning pipelines and a set of well-known machine learning algorithms such as classification, regression, clustering, and collaborative filtering.

- The core concepts the MLlib component introduces to help with building and maintaining complex pipelines are transformer, estimator, and pipeline. A pipeline is an orchestrator that ensures both training and test data flow through identical feature processing steps.

- Model tuning is a critical step in the ML application development process. It is tedious and time-consuming because it involves training and evaluating models over a set of parameter combinations. Combining with the pipeline abstraction, MLlib provides two tools that help: `CrossValidator` and `TrainValidationSplit`.

Managing the Machine Learning Life Cycle

As companies leverage AI and machine learning to transform their business, they soon realize developing and deploying ML applications is not a small task. In Chapter 8, you learned that the machine learning development process is a highly iterative and scientific process that needs an engineering culture and practice that is slightly different from the traditional software development process. As the machine learning development community, including data scientist, *ML engineers and software engineers, gains more experience developing machine learning applications and taking them to production, an apparent theme emerges and has been formalized into a discipline called MLOps.

According to Wikipedia, MLOps is a set of practices that aims to reliably and efficiently develop, deploy, and maintain machine learning models in production. The Google Cloud team defines MLOps as an ML engineering culture and practice that aims to unify ML system development and operation.

As a machine learning pioneer with extensive experience in productionalizing ML applications, Google shared its experience and insights in this area in a seminal paper called "Hidden Technical Debt in Machine Learning Systems" (`https://papers.nips.cc/paper/2015/file/86df7dcfd896fcaf2674f757a2463eba-Paper.pdf`).

This chapter aims to dive deeper into the challenges of developing, managing, and deploying machine learning applications and then shows how an MLflow open source project can help with some of the challenges. In addition, it discusses a few common ML model deployment options.

© Hien Luu 2021
H. Luu, *Beginning Apache Spark 3*, https://doi.org/10.1007/978-1-4842-7383-8_9

The Rise of MLOps

MLOps has become one of the hottest topics among ML practitioners, cloud providers, and startups providing machine learning solutions. Understanding its benefits, best practices, and implementation is real as companies invest in building machine learning applications.

MLOps Overview

MLOps is not a technology or platform. It is an encompassing term for both a set of practices and engineering culture that aim to make developing, deploying, and maintaining, monitoring production machine learning systems seamless, efficient, and reliable. The goal of MLOps is to minimize the technical friction to get the model from an idea to production in the shortest possible time with high-quality predictive power and with as little risk as possible. For many businesses and ML practitioners, only a model running in production can bring value.

At a high level, MLOps advocates for automation and monitoring across the entire ML life cycle to address its unique challenges and needs. Although some of the needs ML systems are similar to the ones in standard software systems, such as continuous integration of source control, unit testing, and continuous delivery, some needs are unique.

- The input data to ML models have a large influence on the quality of ML model predictions; therefore, it is important to test and validate the input data

- Reproducibility. In addition to version control the code to training ML models, additional information must be tracked, such as the input data used for training, the training hyperparameters, and machine learning libraries and their versions.

- ML model quality can degrade easily due to the constant change in the data. Therefore, it is imperative to closely monitor the model performance and machine learning–specific metrics.

To address the machine learning unique challenges, the machine learning community and practitioners have identified a set of best practices for businesses to follow.

- Collaboration

  - Successfully realizing the benefits of machine learning requires collaboration between the various teams within the organization, such as data scientists, ML engineers, data engineers, software engineers, and DevOps engineers. Each team brings unique skills and knowledge to contribute to the various steps in productionalizing machine learning models. Therefore, it requires an engineering culture that promotes and facilitates close collaboration.

- Continuous and Consistent Pipelines

  - Data pipelines that produce the data for the machine learning models to consumers need to be version controlled, run continuously on a certain cadence, and be monitored closely to ensure minimum disruption and high quality.

  - The data pipelines might have specific data transformation logic to produce the features, and therefore that logic needs to be implemented consistently across the training and serving machine learning pipeline.

- Reproducibility

  - Machine learning development is a scientific endeavor that requires iterations, which need reproducibility. It requires iterations to adjust to the constant change in data due to changes such as customer behavior or business objectives. All the assets, artifacts, and metadata that train and evaluate the model must be tracked and version controlled to enable reproducibility.

- Testing and Observability

  - Machine learning models deployment should go through a similar process as standard software deployment, but with some specific validations that are specific to the statistical nature of machine learning, such as the distribution and standard deviations of the model features and model performance evaluation results.

- Once a model is in production to predict new data, it is important to closely monitor and alert on model performance degradation.

Following MLOps best practices enable businesses to dramatically increase the odds of realizing the benefits machine learning provides. As the appetite for leveraging machine learning increases, MLOps make it easier to scale development velocity and maintain many machine learning applications. When the flight wheel of productionalizing machine learning increases in speed, MLOps helps build trust with leadership to reap the benefits through a repeatable process that includes automation, validation, reproducibility, and sound monitoring. Additionally, machine learning increases the return on the investment that businesses have made in building the big data infrastructure over the last decade by capitalizing on the insights from the collected data.

At the time of writing, numerous startups and large cloud providers are racing to invent and build MLOps-related solutions. However, it is difficult to imagine a one-size-fits-all product that can address all MLOps needs.

Companies that have been productionalizing machine learning for a while have something in common: investing in building their own solutions, called a *machine learning platform*. For example, Google has TFX, FB has FBLearner, Uber has Michelangelo, Twitter has Cortex, and LinkedIn has Pro-ML.

The next part of this chapter covers an open source project called MLflow, an open source platform for managing the end-to-end machine learning life cycle.

MLflow Overview

In the MLOps space, there are not many open source projects yet, but I suspect this will change as the ML practitioners and community come together to discuss their needs and learn from each other.

One of the popular open source projects in this area is MLflow, created by Databricks. Its initial release came out in 2018. The capabilities it provides are extremely useful and needed by machine learning practitioners; therefore, its popularity and adoption have steadily increased since the initial release. MLflow's capabilities have been maturing and expanding and becoming more sophisticated as more contributions come from the community.

There are a few reasons behind the MLflow popularity.

- Extensibility

  - MLflow is designed to be open and extensible, so the open source community can easily contribute and extend its core functionalities.

- Flexibility

  - MLflow is designed to work with any ML library and can be used with programming languages in the machine learning community.

- Scalability

  - MLflow is designed to be useful to small and large organizations.

- Run anywhere

  - MLflow can be leveraged and deployed in a company's infrastructure or on most cloud providers or someone's laptop.

One of the reasons behind the success of Apache Spark is its ease of use. Following that recipe, the creators of MLflow wanted to minimize the friction of getting started with MLflow, so they designed MLflow with these two principles in mind: APIs first and modular design. The APIs first principle encourages working backward from the end user's needs and provides a set of programmatic APIs to satisfy those needs. The modular design gives users an easy path to get started and the freedom to incrementally adopt the MLflow platform in a manner that best fits their use cases.

More information on MLflow is at `https://mlflow.org`. The GitHub project is at `https://github.com/mlflow`.

MLflow Components

Logically, the capabilities MLflow platform offers to manage the end-to-end machine learning life cycle can be grouped into four components, as depicted in Figure 9-1. The MLflow components are modular, so you have the flexibility and freedom to adopt one or more components for your machine learning use cases and needs.

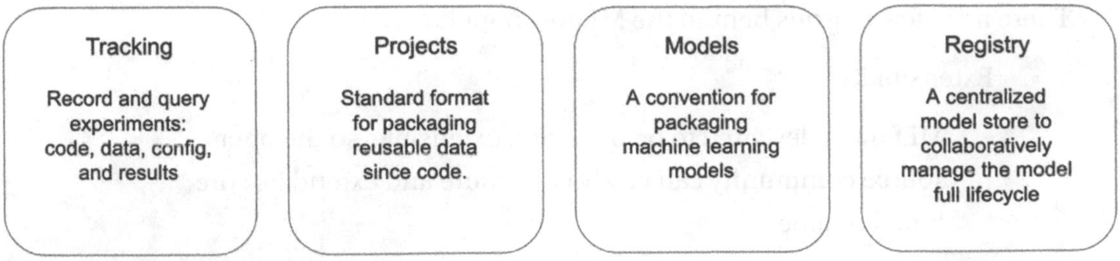

Tracking	Projects	Models	Registry
Record and query experiments: code, data, config, and results	Standard format for packaging reusable data since code.	A convention for packaging machine learning models	A centralized model store to collaboratively manage the model full lifecycle

Figure 9-1. *MLflow components*

As noted, machine learning development is a scientific endeavor that requires running many experiments with various small tweaks on the input to arrive at an optimal model. The Tracking component was designed specifically for this purpose by providing the facilities to track an experiment's inputs, metadata, and output. Once the data of the various runs of an experiment is collected, they can be easily compared, visualized, and shared.

The Projects component defines a standard format for packaging machine learning code as a self-contained executable unit to facilitate the machine learning model reproducibility on various runtime platforms, such as your local laptop or a cloud environment. The ability to easily reproduce machine learning models increases the collaboration between data scientists and their productivity.

The Models component defines a standard format for packaging machine learning models to easily be deployed to various model serving environments, such as on a local machine or cloud providers like Azure, GCP, or AWS.

The Model Registry component provides a centralized way of storing machine learning models to enable a collaborative way of managing their life cycle and lineage. Once a machine learning model is registered, it can help manage model versioning and productionalizing it with an audit trail.

MLflow provides APIs in multiple languages, such as Python, R, Java, and Scala, command-line interface, and UI for you to interact with each of its components.

MLflow in Action

This section goes into more detail about each component to better understand its motivation and learn how to interact with them using the provided UI, command-line tools, and APIs.

The prerequisites for running the following examples are Python 3.x, scikit-learn 0.24.2, and MLflow 1.18 or higher. Assuming your computer already has Python 3.x installed, you can install the remaining ones using the commands listed in Listing 9-1.

Listing 9-1. Installing MLflow and scikit-learn

```
pip install scikit-learn
pip install mlflow

mlflow --version    # to test the installation and version
# you should see the output similar to below
mlflow, version 1.18.0
```

MLflow Tracking

The motivation behind the MLflow Tracking component is to enable data scientists to track all the artifacts that are needed and produced while developing and optimizing their models during the model development phase. During this phase, data scientists usually need to run many experiments to optimize their model performance by tweaking the various input parameters, such as the input features, algorithm, and hyperparameters.

Traditionally, data scientists track the details of their experiments using either a notepad, a document, or spreadsheet. Unfortunately, this approach is manual and error-prone, and the experiment results are not easily shared, visualized, and compared with other experiments.

Conceptually, the tracking information produced each time the model training code runs is organized into a *run,* and you can record the following information.

- **Parameters**: Key/value input parameters of your choice, where the value is a string. Examples of parameters are hyperparameters like learning rate, number of trees, and regularization.

- **Metrics**: Key/value metrics, where the value is numeric. Each metric can be updated throughout a run. You can visualize its full history. Examples of metrics include accuracy, RMSE, and F1.

- **Artifacts**: Output files in any format. Examples of artifacts are the image of the accuracy and feature importance.

- **Tags**: One or more key/value tags in the currently active run.

- **Metadata**: General information about the run, such as run date and time, name, source code, and code version.

You can optionally organize multiple runs into an experiment, which is usually designed for a specific machine learning task. MLflow Tracking component provides APIs to record the run, and they are available in multiple languages: Python, R, Java, Scala, and REST APIs. The tracking server records the run information, which can run locally on the same machine as the application calling the APIs or on a remote machine. Essentially, MLflow tracking is a client-server application with the architecture depicted in Figure 9-2.

For learning purposes or exploration, it is easier to run the ML tracking server locally. In a team environment in which your team wants to centrally track and manage a machine learning model life cycle, it makes more sense to configure and run the tracking server on a remote host. MLflow provides two easy ways to specify where the tracking server is running on. The first way is by setting the MLFLOW_TRACKING_URI environment variable with a URI of the tracking server. The second way is to specify such URI using the mlflow.set_tracking_uri() in your application. If the tracking server URI is not set, the tracking APIs log the run information locally in a mlruns directory wherever you run your program.

Once the run information is available in the server, you can access it via the provided MLflow tracking UI, use the tracking APIs, or use Spark, as depicted on the right part of Figure 9-2.

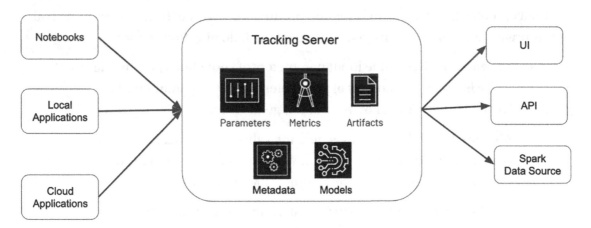

Figure 9-2. *MLflow tracking component architecture*

The run information can be categorized into two types: structured data and unstructured data. The artifacts, such as images, models, or data files, are considered unstructured and stored in the artifact store. The rest of the run information is considered structured data and is stored in the back-end store. The artifact store can be a folder on a local file system or one of the cloud providers' distributed storage, such as AWS S3, Azure blob storage, or Google Cloud storage. The back-end store can be a folder on a local file system or one of the SQL stores, such as Postgres, MySQL, or SQLite, as depicted in Figure 9-3.

In terms of storing the artifacts, it turns out the MLflow client APIs get the artifact store URI from the tracking server. Then it is responsible for uploading the artifacts directly to the artifact store.

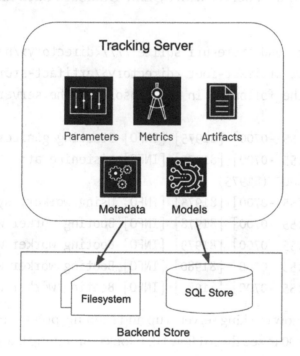

Figure 9-3. *MLflow tracking server back-end store options*

With a good understanding of the MLflow tracking component architecture and where the run information is stored, the next part shows how to launch an instance of an MLflow tracking server, use the tracking APIs to track the run, and then use the tracking UI to visualize the run information.

Listing 9-2 uses a local directory as an artifact store and a local SQLite file as the back-end store. Before launching the MLflow server to manage the run information and the artifacts, you need to create two directories: one is for the database-backend store, and the other one is for the artifact store. Listing 9-2 shows the command to start the MLflow server.

Note   In order to use model registry functionality, you must run your MLflow tracking server using a database-backend store

---

Listing 9-2. Start MLflow Server with SQLlite Database Back End and Local Artifact Store

```
mlflow server --backend-store-uri sqlite:////<directory>/backend-store/
mlflow.db --default-artifact-root <directory>/artifact-store
# you should see the following in the console if the server was started
successfully
[2021-07-24 08:17:55 -0700] [81975] [INFO] Starting gunicorn 20.0.4
[2021-07-24 08:17:55 -0700] [81975] [INFO] Listening at:
http://127.0.0.1:5000 (81975)
[2021-07-24 08:17:55 -0700] [81975] [INFO] Using worker: sync
[2021-07-24 08:17:55 -0700] [81978] [INFO] Booting worker with pid: 81978
[2021-07-24 08:17:55 -0700] [81979] [INFO] Booting worker with pid: 81979
[2021-07-24 08:17:55 -0700] [81980] [INFO] Booting worker with pid: 81980
[2021-07-24 08:17:55 -0700] [81981] [INFO] Booting worker with pid: 81981
```

Now that the MLflow tracking server is up and running, point your browser at http://localhost:500 to see the MLflow UI. It looks something like in Figure 9-4.

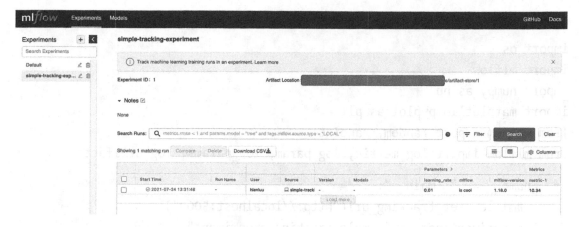

Figure 9-4. *MLflow UI*

The first example demonstrates the usage of the various tracking APIs to track the different pieces of information of a run. The source code of this example is in the chapter9/simple-tracking.py file. By default, this example sets the MLflow tracking URI to http://localhost:5000, so before you execute this Python script using the command in Listing 9-3, make sure your MLflow tracking server is already up and running.

Listing 9-3. Executing simple-tracking.py

```
python simple-tracking.py

# the output would looking something like below
starting a run with experiment_id 1
done logging artifact
Done tracking on run
experiment_id: 1
run_id: cb17324d40764a428b3d983e8ac4d1dd
```

The simple-tracking.py script shown in Listing 9-4 is written in a safe way to run multiple times, and each time it creates a new run under the same experiment called simple-tracking-experiment. If you run it five times, the MLflow tracking UI would look something like Figure 9-5, which shows the runs in a table format with information such as the start time of each run, the user who logs the run information, and so on.

Listing 9-4. Content of simple-tracking.py

```python
import os
import mlflow
import numpy as np
import matplotlib.pyplot as plt
from random import random, randint
from mlflow import log_metric, log_param, log_params, log_artifacts

if __name__ == "__main__":
    #mlflow.set_tracking_uri("http://localhost:5000")
    experiment_name = "simple-tracking-experiment"
    experiment = mlflow.get_experiment_by_name(experiment_name)
    experiment_id = experiment.experiment_id if experiment else None
    if experiment_id is None:
    print("INFO: '{}' does not exist. Creating a new experiment
                    experiment".format(experiment_name))
    experiment_id = mlflow.create_experiment(experiment_name)

    print("starting a run with experiment_id
                        {}".format(experiment_id))
    with mlflow.start_run(experiment_id=experiment_id) as run:
    # Log a parameter (key-value pair)
    log_param("mlflow", "is cool")
    log_param("mlflow-version", mlflow.version.VERSION)

    params = {"learning_rate": 0.01, "n_estimators": 10}
    log_params(params)

    # Log a metric; metrics can be updated throughout the run
    log_metric("metric-1", random())
    for x in range(1,11):
        log_metric("metric-1", random() + x)

    # Log an artifact (output file)
    if os.path.exists("images"):
        log_artifacts("images")
        print("done logging artifact")
```

```
else:
    print("images directory does not exists")

image = np.random.randint(0, 256, size=(100, 100, 3),
                          dtype=np.uint8)
mlflow.log_image(image, "random-image.png")

fig, ax = plt.subplots()
ax.plot([0, 2], [2, 5])
mlflow.log_figure(fig, "figure.png")

experiment = mlflow.get_experiment(experiment_id)
print("Done tracking on run")
print("experiment_id: {}".format(experiment.experiment_id))
print("run_id: {}".format(run.info.run_id))
```

Figure 9-5. *MLflow tracking UI after five runs*

One very useful feature in MLflow UI is to compare the metrics across multiple runs. You simply select two or more runs by clicking the check box of those runs and then click the Compare button to compare them side by side, as depicted in Figure 9-6.

407

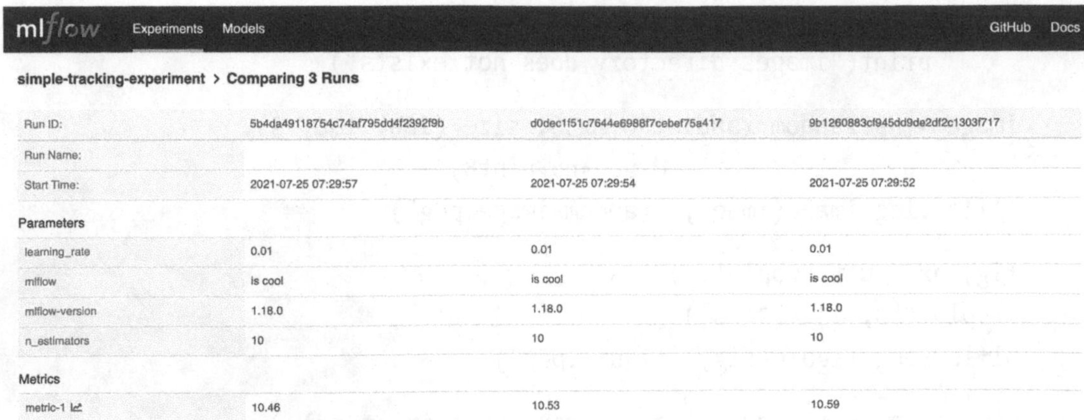

Figure 9-6. *Comparing runs side by side*

Let's take a closer look at what's going on in the `simple-tracking.py` script. It first sets the tracking URI of the tracking server. Then it determines whether to create an experiment called simple-tracking-experiment, if it doesn't already exist. Next, it uses MLflow, a high-level fluent API, to start a run using the Python `with block` and automatically terminate the run at the end of the `with` block. The `with` block contains examples of various tracking APIs to log parameters, metrics, artifacts, images, and figures. I highly recommend you visit the MLflow API documentation website, such as `https://mlflow.org/docs/latest/python_api/index.html`, to learn about the various APIs and their usage.

If you comment out the line that sets tracking URI and runs the command in Listing 9-3, MLflow tracking APIs log the run information in a local directory called `mlruns`. To view the run information in that local directory, run the `mlflow ui` command and then point your browser to `http://localhost:5000`.

Each run belongs to a specific experiment, so when starting a run without specifying an experiment ID, MLflow creates it in the default experiment.

To examine the detailed information about each run, simply click the link of the run start time. You see something like Figure 9-7.

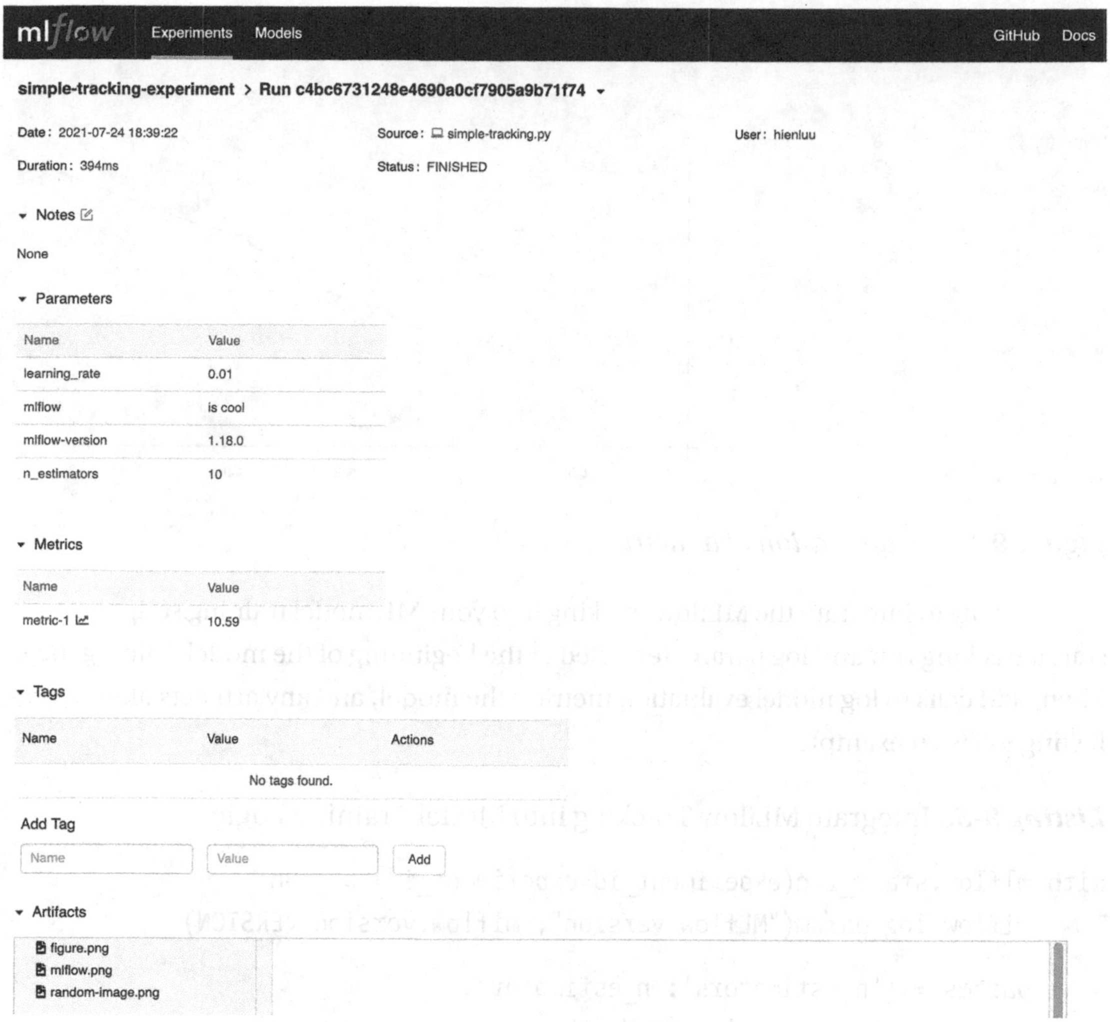

mlflow   Experiments   Models   GitHub   Docs

simple-tracking-experiment > **Run c4bc6731248e4690a0cf7905a9b71f74** ▾

Date : 2021-07-24 18:39:22   Source : ▢ simple-tracking.py   User : hienluu

Duration : 394ms   Status : FINISHED

▾ Notes ☑

None

▾ Parameters

Name	Value
learning_rate	0.01
mlflow	is cool
mlflow-version	1.18.0
n_estimators	10

▾ Metrics

Name	Value
metric-1 �📈	10.59

▾ Tags

Name	Value	Actions
	No tags found.	

Add Tag

Name	Value	Add

▾ Artifacts

📄 figure.png
📄 mlflow.png
📄 random-image.png

Figure 9-7. *Detailed information of a run*

The general metadata about a run is displayed at the top, and then each type of
tracking information is displayed in separate sections. For each metric that was updated
multiple times, you can see the visualization of each metric by clicking the link of the
metric name. A graph of metric-1 is depicted in Figure 9-8.

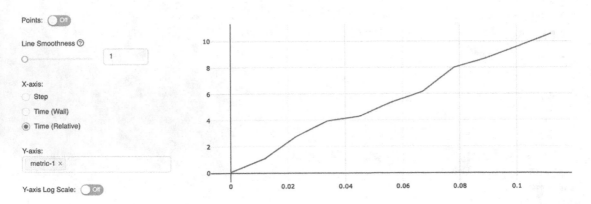

Figure 9-8. *Visualization of a metric*

One way to integrate the MLflow tracking into your ML model training script is to start a tracking run and log parameters used at the beginning of the model training logic. Then, add calls to log model evaluation metrics, the model, and any artifacts after it. Listing 9-5 is an example.

Listing 9-5. Integrate MLflow Tracking into Model Training Logic

```
with mlflow.start_run(experiment_id=experiment_id) as run:
    mlflow.log_param("MLflow version", mlflow.version.VERSION)

    params = {'n_estimators': n_estimators,
              'max_depth': max_depth,
              'min_samples_split': min_samples_split,
              'learning_rate': learning_rate, 'loss': 'ls'}
    mlflow.log_params(params)

    gbr = ensemble.GradientBoostingRegressor(**params)
    gbr.fit(X_train, y_train)

    y_pred = gbr.predict(X_test)
    # calculate error metrics
    mae = metrics.mean_absolute_error(y_test, y_pred)
    mse = metrics.mean_squared_error(y_test, y_pred)
```

```
rsme = np.sqrt(mse)
r2 = metrics.r2_score(y_test, y_pred)

# Log model
mlflow.sklearn.log_model(gbr, "GradientBoostingRegressor")

# Log metrics
mlflow.log_metric("mae", mae)
mlflow.log_metric("mse", mse)
mlflow.log_metric("rsme", rsme)
mlflow.log_metric("r2", r2)

experiment = mlflow.get_experiment(experiment_id)
print("Done training model")
print("experiment_id: {}".format(experiment.experiment_id))
print("run_id: {}".format(run.info.run_id))
```

As it turns out, the need for logging metrics, parameters, and models are common when training machine learning models using the various machine learning libraries, such as scikit-learn, TensorFlow, Spark, and Keras. The MLflow Tracking component goes one step further to simplify this process by providing an API called `mlflow.autlog()`. Adding this line of code before your model training code automatically logs all the common information without the need for explicit log statements. Listing 9-6 is an example.

Listing 9-6. MLflow Automatic Logging

```
# enable auto logging
mlflow.autolog()

# prepare training data
X = np.array([[1, 1], [1, 2], [2, 2], [2, 3]])
y = np.dot(X, np.array([1, 2])) + 3

# train a model
model = LinearRegression()
with mlflow.start_run() as run:
    model.fit(X, y)
```

At the time of writing this book, support for autologging was in an experimental state. Please consult the documentation at `https://mlflow.org/docs/latest/tracking.html#automatic-logging` for the latest information on each supported library.

MLflow Projects

The MLflow Projects component standardizes the project packaging format to be reusable and reproducible on multiple platforms.

Several innovations around machine learning libraries train the models, such as TensorFlow, PyTorch, Spark MLlib, and XGBoost. Data scientists tend to favor the library that can help them produce optimized machine learning models for their business use cases. Nowadays, numerous computing resources are available to data scientists to train small to large machine learning models, such as local machines, Docker, on the cloud, and so forth.

The Projects component organizes and defines machine learning projects to capture code, configuration, dependencies, and data in an executable unit. As a result, data scientists can easily use any machine learning library in their projects and run their projects on any computing platform (see Figure 9-9).

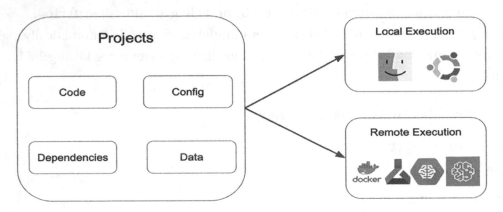

Figure 9-9. *MLFlow project details*

Each MLflow project is simply a directory of files or a Git repository. Although it is optional, it is highly recommended that your project contains a file called MLproject, which specifies the environment, parameters, and entry points to control the execution of your project. A project supports the following type of environments, and each one requires its own way of defining it.

- **Conda**: Uses the Conda package management system, which can support Python packages and native libraries, execute your MLflow project on

- **Docker**: Uses a Docker container environment that can support almost any type of dependencies to execute your MLflow project on

- **System**: Your current system environment to execute your MLflow project on

For more information about the various supported environments in MLflow, refer to the MLflow projects documentation at `https://mlflow.org/docs/latest/projects.html#specifying-projects`.

In addition to the MLproject file, an MLflow project usually includes a file to define the environment and another one that contains the model training logic. Listing 9-7 shows the content of a sample MLproject file and `conda.yml` file in an MLflow project that uses the Conda environment. It is a good practice to set up the parameters in the MLproject like Listing 9-7, so they can be easily overwritten from the command line so data scientists can easily try out different values in their model optimization process.

Listing 9-7. An Example of MLproject File with Conda Environment

```
# MLproject file
name: boston-housing-price

conda_env: conda.yaml

entry_points:
  main:
    parameters:
      run_name: {type: str, default: "run_name"}
      n_estimators: {type: int, default: 100}
      max_depth: {type: int, default: 4}
      min_samples_split: {type: int, default: 2}
      learning_rate: {type: float, default: 0.01}
    command: |
      python train.py \
```

```
      --n_estimators={n_estimators} \
      --max_depth={max_depth} \
      --min_samples_split={min_samples_split} \
      --learning_rate={learning_rate}

# conda.yaml
channels:
- conda-forge
dependencies:
- python=3.7.6
- pip
- pip:
  - mlflow
  - scikit-learn==0.24.2
  - cloudpickle==1.6.0
```

Now that you know how to put together an MLflow project, the next part is learning how to run them. MLflow Projects component provides two ways to run projects programmatically: the `mlflow run` command-line tool and the `mlflow.projects.run()` API. Both ways take similar parameters and work similarly. Listing 9-8 runs an MLflow project using the command-line tool. You can display its usage by issuing the `mlflow run --help` command.

The first and important parameter is the project URI, which is either a directory on the local file system or a Git repository path. Listing 9-8 contains several examples of running an MLflow project that exists in a local directory.

Listing 9-8. Run MLflow Project from Local Directory

```
# run the boston-housing-price MLflow project with creating a
# new conda environment and using default parameter values and
# add run under the boston-housing-price experiment.

mlflow run <chapter9>/boston-housing-price --experiment-name=boston-
housing-price
```

```
# similar to the one above, except without creating a
# new conda environment

mlflow run <chapter9>/boston-housing-price --no-conda --experiment-
name=boston-housing-price

# to overwrite one or more parameter value, specify them using -P # format
mlflow run <chapter9>/boston-housing-price --no-conda -P learning_rate=0.06
--experiment-name=boston-housing-price
```

When running an MLflow project that uses a Conda environment, MLflow first creates a new Conda environment and then downloads all dependencies specified in the conda.yaml file, therefore it might take a while to complete all the steps. This is useful when trying to reproduce the model from someone else's MLflow project or validate an MLflow project's reproducibility. Specifying the --no-coda command parameter skips the Conda creation step, which speeds up the project building process. This is very useful when you are putting together your MLproject.

To accommodate the various application development infrastructures, MLflow projects support other environments such as Docker and Kubernetes. They provide more flexibility but are a bit more complex to set up.

MLflow Models

The motivation behind MLflow Models component is to promote model interoperability by standardizing the ML model packaging format so that they can be developed using any of the popular machine learning libraries and deployed to a diverse set of execution environments, as depicted in Figure 9-10. For example, you could develop a model in PyTorch, and deploy it and perform inference on your local Docker, Spark, or one of the cloud provider ML platforms. The solution MLflow Models uses is by defining a unified model abstraction that captures the flavor of the model.

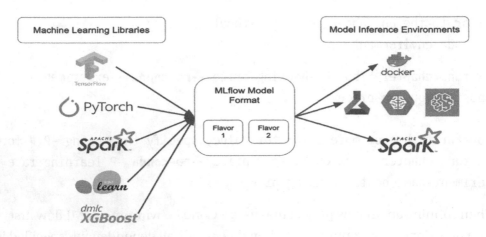

Figure 9-10. *MLflow model abstraction*

Flavors are the key concept that makes the MLflow Models component versatile and useful. Essentially, flavors are a convention that deployment tools can decipher to understand the model; therefore, it is possible to develop tools that work with models trained using any ML library without integrating each tool with each specific library. Out of the box, MLflow defines several supported flavors that all its built-in deployment tools support.

Similar to an MLflow project, an MLflow model is a directory containing a set of files. Among them is a file called MLmodel, which contains a few pieces of metadata about the model and defines the flavors that the model can be viewed in. If your model training script logs the model using API `log_model` or saves a model using API `save_model`, a model directory is automatically created with all the appropriate files that contain information about the environment and dependencies to load and serve it. Figure 9-10 shows an example of an MLflow model directory and its content from the boston-housing-price project. Navigate to one of the runs under the boston-housing-price experiment and then scroll down to the artifacts section. You see something like Figure 9-11.

▼ Artifacts

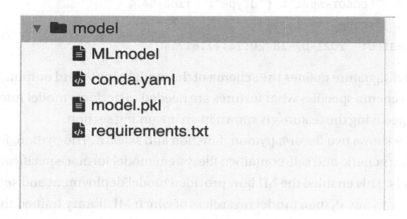

Figure 9-11. *MLFlow model directory and its content*

The MLmodel file captures some of the model's metadata, such as when it was created and run. More importantly, it also contains the model signature and flavors. Listing 9-9 shows the content of MLmodel file generated by the mlflow.autoLog API in the chapter9/airbnb-price/train.py.

Listing 9-9. Content of MLmodel

```
artifact_path: model
flavors:
  python_function:
    env: conda.yaml
    loader_module: mlflow.sklearn
    model_path: model.pkl
    python_version: 3.7.6
  sklearn:
    pickled_model: model.pkl
    serialization_format: cloudpickle
    sklearn_version: 0.24.2
run_id: fc9bf6efeff74752812debc131b6c369
signature:
  inputs: '[{"name": "bedrooms", "type": "double"},
          {"name": "beds", "type": "double"},
            {"name": "bathrooms", "type": "double"}]'
```

417

```
outputs: '[{"type": "tensor",
            "tensor-spec": {"dtype": "float64",
            "shape": [-1]}}]'
utc_time_created: '2021-07-28 20:14:17.612140'
```

The model signature defines the schema of the model's input and output. The model input schema specifies what features are needed to perform model inference. An example of specifying the features is shown in an upcoming section.

Listing 9-9 shows two flavors: python_function and sklearn. The python_function flavor defines a generic and self-contained filesystem model format, specifically for Python models. This enabled the MLflow provided model deployment and serving tools to work with any Python model regardless of which ML library trained the model. As a result, any Python model can be easily productionalized in a variety of runtime environments.

`conday.yaml` and `requirements.txt` capture the dependencies and environment information, respectively, so a similar environment can easily be created at the deployment time.

The MLflow built-in model persistence utilities take care of packaging models for the various popular ML libraries, such as PyTorch, TensorFlow, scikit-learn, LightGBM, and XGBoost. If your model requires special handling, MLflow supports persisting and loading custom model format.

In addition to providing a set of APIs to manage the model life cycle, MLflow's Models component provides command-line tools to deploy, load, and serve models.

To demonstrate the usage of command-line tools, the next section uses the `airbnb-price` MLflow project, which is a simple MLflow project to predict the price of an Airbnb listing using the scikit-learn Random Forest algorithm. For simplicity's sake, it uses only three features: number of bedrooms, number of beds, and number of bathrooms. This project is located in the `chapter9/airbnb-price` folder, and the train.py training script uses the `mlflow.autoLog` API to automatically log the parameters, metrics, and model. To train the model, you can issue one of the commands listed in Listing 9-10. This example assumes the MLflow has already started and is running on port 5000 on your local machine.

Listing 9-10. Run airbnb-price MLflow Project

```
# make sure to set the MLFLOW tracking server URI first
export MLFLOW_TRACKING_URI=http://localhost:5000

# run airbnb-price project
# with the default 100 estimators and max depth of 4
mlflow run ./airbnb-price --no-conda  --experiment-name=airbnb-price
# with the 300 estimators and max depth of 9
mlflow run ./airbnb-price --no-conda  --experiment-name=airbnb-price -P
n_estimators=300 -P max_depth=9
```

Next, navigate to the latest run logged under the airbnb-price experiment in MLflow Tracking UI, and locate run_id in the MLmodel file under the Artifacts section. Next, you use the mlflow serve command-line tool to serve the model associated with the provided run id by launching a web server on your local machine. The mlflow serve command in Listing 9-11 launches a web server running with port 7000 and instructs MLflow to use the python_function flavor.

Listing 9-11. Launch Webserver to Perform Model Inference

```
# replace run id with the real run id
# the command below will launch the webserver that
# listens on port 7000.

mlflow models serve --model-uri runs:/<run id>/model -p 7000 --no-conda

# the output of the above command looks something like below
2021/07/28 19:50:25 INFO mlflow.models.cli: Selected backend for flavor
'python_function'
2021/07/28 19:50:25 INFO mlflow.pyfunc.backend: === Running command
'gunicorn --timeout=60 -b 127.0.0.1:7000 -w 1 ${GUNICORN_CMD_ARGS} --
mlflow.pyfunc.scoring_server.wsgi:app'
[2021-07-28 19:50:26 -0700] [36709] [INFO] Starting gunicorn 20.0.4
[2021-07-28 19:50:26 -0700] [36709] [INFO] Listening at:
http://127.0.0.1:7000 (36709)
[2021-07-28 19:50:26 -0700] [36709] [INFO] Using worker: sync
[2021-07-28 19:50:26 -0700] [36712] [INFO] Booting worker with pid: 36712
[2021-07-28 19:54:24 -0700] [36709] [INFO] Handling signal: winch
```

To perform inference using the `airbnb-price` Random Forest model, you send HTTP requests using the `curl` command-line tool to the `invocations` REST endpoint. Listing 9-12 contains a few examples to predict Airbnb listing prices.

Listing 9-12. Perform Model Inferencing Using HTTP Requests

```
# single prediction
curl http://127.0.0.1:7000/invocations -H 'Content-Type: application/json'
-d '{"columns": ["bedrooms","beds","bathrooms"], "data": [[1,1,1]]}'
# multiple predictions
curl http://127.0.0.1:7000/invocations -H 'Content-Type: application/json'
-d '{"columns": ["bedrooms","beds","bathrooms"], "data": [[1,1,1], [2,2,1],
[2,2,2], [3,2,2]]}'

# The HTTP request response contains a single value, which is the predicted
price of an Airbnb listing with the specified features.
```

You can also perform the model inference programmatically using the predicted API in the `mlflow.model` module.

The MLflow Model component provides many other capabilities. More information is at `www.mlflow.org/docs/latest/models.html`.

MLflow Model Registry

The motivation behind the MLflow Registry component is to provide means for managing the complete life cycle of an MLflow model, as depicted in Figure 9-12. This life cycle consists of the lineage information about the MLflow experiment and runs that produced the model, the model registration and versioning, and a workflow to transition the model from one stage to another in the deployment process with audit trail and notes. This component is the most recent and was introduced in MLflow 1.7.

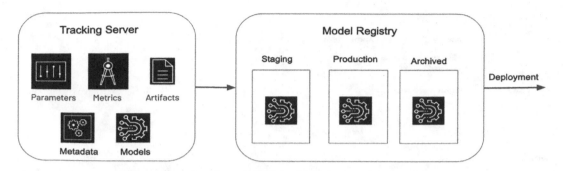

Figure 9-12. *MLflow model life cycle*

Note   To use model registry functionality, you must run your MLflow tracking server using a database back-end store.

Like the other MLflow components, this one also provides UI, APIs, and command-line tools for you to interact with. The next section provides examples of managing the model's life cycle produced from one of the runs in experiment `airbnb-price`.

The first step in the MLflow model life cycle is model registration. Before you can add an MLflow model to the Model Registry, you must log in using `log_model` API or via the `autolog` API. Each registered model can have one or more versions. This model name and version combination makes it easy to perform inference and track A/B testing before fully launching it to production. When a model is registered with the Model Registry, a name must be provided. If the model name doesn't already exist, then it is added as of version 1. Otherwise, a new model version is automatically created.

To register a model produced by a certain run from the UI, you first navigate to the detail page of the run, scroll down to the Artifacts section, select the top-level folder, and click the Register Model button (see Figure 9-13).

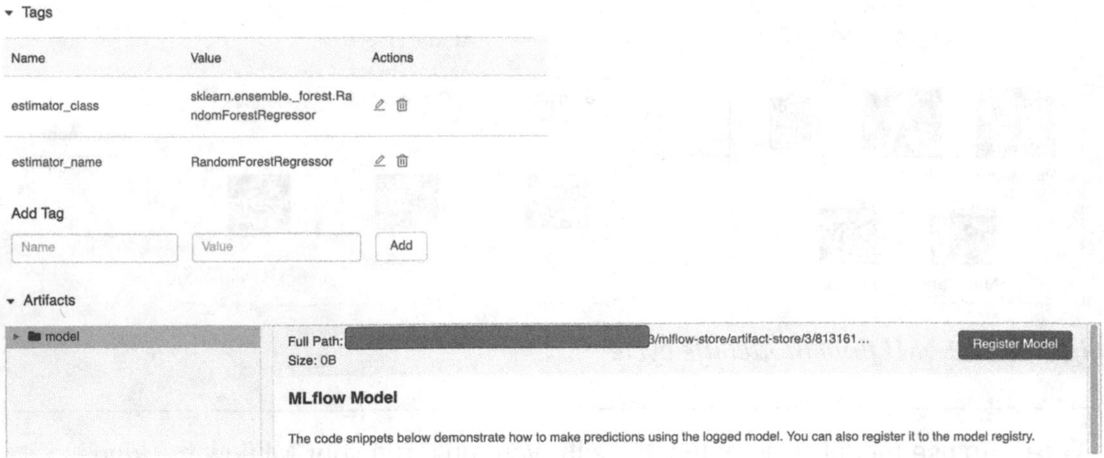

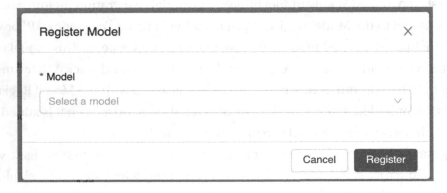

Figure 9-13. *Model registration*

The Register Model dialog box pops up for you to enter a model name, as depicted in Figure 9-14. If the model name already existed, then you see a drop-down list for you to select.

Register Model   ✕

* Model

Select a model   ⌄

Cancel   Register

Figure 9-14. *Register model dialog box*

To view the registered model once the model registration is completed, navigate to the Registered Models page to see all the registered models by clicking the Models link at the top of MLflow UI. You see something like Figure 9-15.

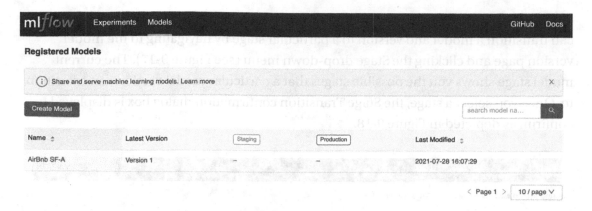

Figure 9-15. *Registered model list page*

Each model has an overview page to show the various active model versions. To see the Airbnb SF-A model's overview page, click the model name. You see something like Figure 9-16.

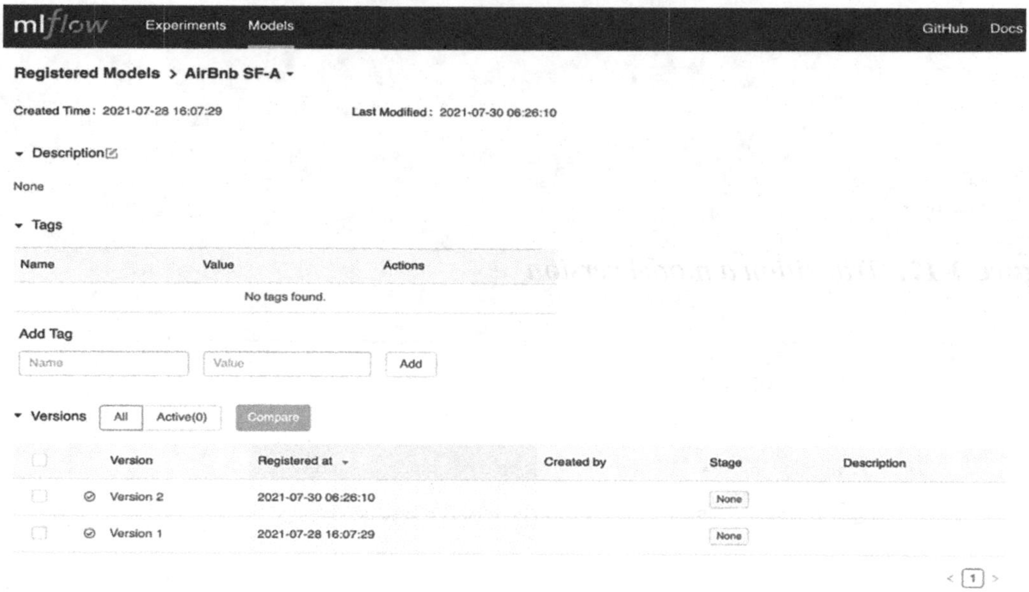

Figure 9-16. *Registered model details page*

The built-in stages of a registered model are *staging, production,* and *archived.* You can transition a model and version to a particular stage by navigating to the model version page and clicking the Stage drop-down menu (see Figure 9-17). The current model stage shows you the possible stages that a particular model version can transition to. Once you select a stage, the Stage Transition confirmation dialog box is displayed to confirm, as depicted in Figure 9-18.

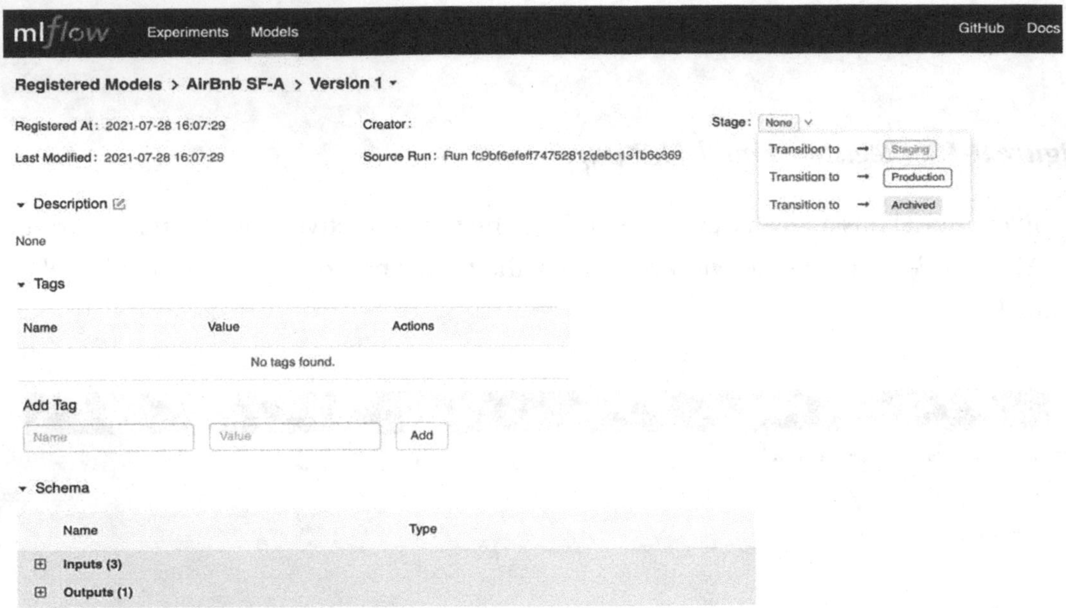

Figure 9-17. *Transition a model version*

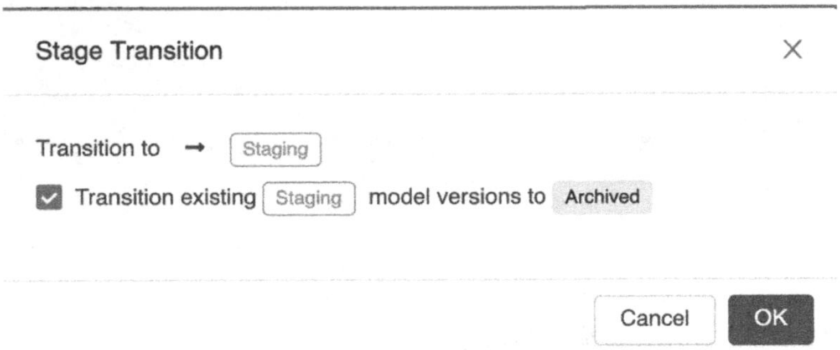

Figure 9-18. *Model stage transition*

If a registered model has multiple versions and is in different stages, the model overview page gives you a great bird's eye view of what's going on. An example of this is depicted in Figure 9-19.

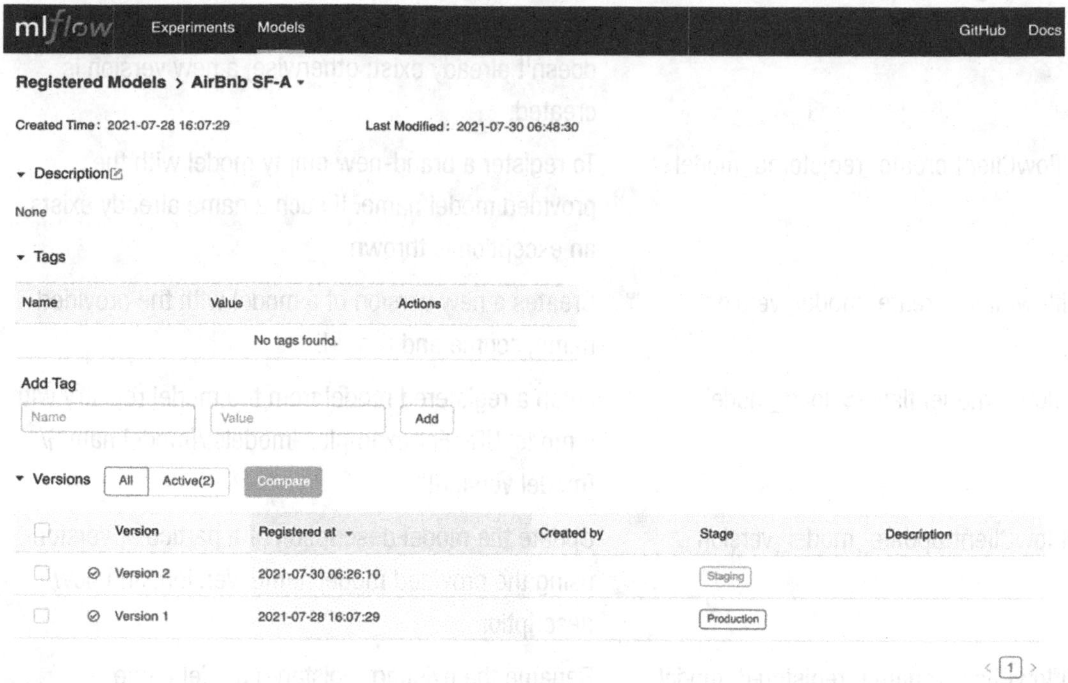

Figure 9-19. *Bird's-eye view of the stage of model versions*

The preceding examples use MLflow Model Registry UI to manage the life cycle of models, from registration to transition them to various stages. You can programmatically perform the same tasks by using the provided APIs. The APIs, listed in Table 9-1, make it easy to integrate the model management life cycle with a CI/CD system. For example, if a model is trained continuously by a CI/CD pipeline at a regular cadence, and if the model performance passes the predetermined criteria, the CI/CD pipeline can easily transition to the next appropriate stage for data scientists to analyze and determine the next step.

Table 9-1. *APIs to Interact with Model Registry*

Name	Description
mlflow.register_model	Add a model to the registry using run URI and model name. Version 1 is created if the provide model name doesn't already exist; otherwise, a new version is created.
MlflowClient.create_registered_model	To register a brand-new empty model with the provided model name. If such a name already exists, an exception is thrown.
MlflowClient.create_model_version	Creates a new version of a model with the provided name, source and run_id.
mlflow.<model flavor>.load_model	Fetch a registered model from the model registry with a model URI. For example, "models:/{model name}/{model version}"
MlflowClient.update_model_version	Update the model description of a particular version using the provided model name, version, and new description.
MlflowClient.rename_registered_model	Rename the existing registered model name.
MlflowClient.transition_model_version_stage	Transition a registered model to one of the stages: staging, production, or archived
MlflowClient.list_registered_models	Fetch all the registered models in the registry.
MlflowClient.search_model_versions	Search for a list of model versions using a registered model name.
MlflowClient.delete_model_version	Delete a specific version of a registered model name.
MlflowClient.delete_registered_model_version	Delete a registered model and all its versions.

For a comprehensive list of Model Registry APIs, please read the MLflow Model Registry API workflow documentation at https://mlflow.org/docs/latest/model-registry.html#api-workflow.

Model Deployment and Prediction

The model deployment strategy is largely dependent on the model prediction needs, and these two tend to go hand in hand. Different machine learning use cases have different needs and requirements when it comes to the model prediction. Some of the standard requirements are latency, throughput, and cost. Up until recent times, the model deployment topic is usually left out of machine learning research papers.

When applying machine learning to business use cases, it is important to understand the different deployment options and when to use them. This section describes two common model deployment strategies and model prediction scenarios and where Spark can fit in.

The two common model prediction scenarios are online prediction and offline prediction. Table 9-2 compares these two scenarios in terms of the standard requirements.

Table 9-2. *Online Prediction vs. Offline Prediction*

Scenario	Latency	Throughput	Cost
Online	Milliseconds	Low	Vary
Offline	Seconds to days	High	Vary

Online prediction scenario is used when machine learning model prediction is a part of an online system to perform a certain user activity, which usually means it needs to be fast. Therefore, the latency needs to be in milliseconds. Examples of online predictions are online advertisement, fraud detection, search and recommendation, and many more. The deployment strategy for online prediction involves building and managing a prediction service that supports REST or gRPC protocol to perform model predictions concurrently and support at a high request rate. The response latency must be low—in tens of milliseconds. In other words, the prediction service must be scalable and reliable. The prediction service usually is a stateless service that sits behind a load balancer and runs on a cluster of machines or Kubernetes nodes. The cost associated with online prediction is a function of the latency requirement and the scale at which the prediction request rate will be.

The model prediction provided by the Spark MLlib component cannot meet the low latency requirement. Therefore, you either train your machine learning models using a machine learning library such as PyTorch, TensorFlow, and XGBoost or export your MLlib trained model outside of Spark using the tools from ONNX (http://onnx.ai).

The offline prediction scenario is very useful when the model prediction needs to be performed in large batches at a certain cadence. It is not an integral part of the online user flow. Examples of offline predictions are movie recommendations generated per user, user churn propensity predictions, market demand forecasting, and customer segmentation analysis. These offline predictions are usually written out to a persistent distributed storage or low latency distributed database for downstream systems to access the predictions or serve online user traffic. This is the easiest and cheapest deployment strategy in terms of complexity and cost because the offline predictions are made using batch jobs, and the cost is comparatively low due to low overhead. It is incurred only while those jobs are running.

Spark is a great choice for this scenario due to its scalable and distributed computing framework, well-integrated MLlib component for model training and evaluation, and the ease of integration between the MLflow Model Registry component for model life cycle management and batch jobs. One important consideration in offline prediction is the frequency of generating predictions. The answer depends on how important the prediction freshness is to your machine learning use case. For the movie recommendation use case, probably the closer to real-time, the better, but maybe it is sufficient that the frequency can be in hours. One small optimization the offline prediction can make for this use case is to skip generating recommendations for users that haven't been active in the last few months.

Summary

- MLOps brings best practices and an engineering mindset to productionalizing machine learning applications so businesses worldwide can reap the benefits machine learning brings to business use cases.

- MLflow is an open source platform for managing the machine learning life cycle. It provides four components to help with steps in the machine learning development process

- The Tracking component enables data scientists to track all the artifacts needed and produced while developing and optimizing their model during the model development phase.

- The Projects component standardizes the packaging format of machine learning projects to be reusable and reproducible on multiple platforms.

- The Models component standardizes the packaging format of machine learning models to be developed using any popular machine learning libraries and be deployed to a diverse set of execution environments.

- The Model Registry component provides a mechanism to manage model life cycle and lineage using a central repository to host the registered models, a workflow to transition models through its life cycle, and UI and APIs to interact with registered models.

- Model deployment and prediction go hand in hand. The two common model prediction scenarios are online and offline, and each is appropriate for a different set of use cases.

Index

A

Adaptive Query Execution (AQE)
    framework
    definition, 204
    features, 206
    shuffle partitions, 206–208, 210, 211
    skew joins, 213–218
    Spark SQL Catalyst, 205
    switching join strategies, 211–213
    transformations, 205
Advanced analytics functions, 160
agg function, 124, 126
Aggregations, 111
    average value, 118
    collection group values, 124
    counting, 114, 116
    dataset, 113, 114
    functions, 112
    grouping, 121, 122
    minimum/maximum values, 117
    multiple, 123
    pivoting, 125–127
    skewness, 118, 119
    standard deviation, 120, 121
    sum, 117
    sum up, 118
    variance, 120, 121
AlphGo, 333
ALSModel class, 384, 387

Alternate-least-square, 383
Analytic functions, 168
Apache Apex, 229, 383
Apache Beam, 229, 230
Apache Flink, 229, 284
Apache Kafka, 229
Apache Samza, 229
Apache Spark applications, 11
Apache Storm, 229
append output mode, 275, 276
approx._count_distinct
    function, 116, 143
Arbitrary stateful processing
    cases, 304
    flatMapGroupsWithState
        API, 310–315
    mapGroupsWithState, 305–307, 309
    state timeouts, 303, 304
    structure streaming, 300, 301
Artificial intelligence (AI), 331
avg function, 171, 293
awaitTermination() function, 233, 249

B

Batch data processing, 221–223, 230
Bell-shaped curve distribution, 365
Big data applications, 6, 14
Binary classification, 335, 376
BinaryClassificationEvaluator, 377, 379

© Hien Luu 2021
H. Luu, *Beginning Apache Spark 3*, https://doi.org/10.1007/978-1-4842-7383-8

Printed in the United States
by Baker & Taylor Publisher Services